T0198458

Transformative AI
Responsible, Transparent, and Trustworthy AI Systems

RIVER PUBLISHERS SERIES IN COMPUTING AND INFORMATION SCIENCE AND TECHNOLOGY

Series Editors:

K.C. CHEN
National Taiwan University, Taipei, Taiwan
University of South Florida, USA

SANDEEP SHUKLA
Virginia Tech, USA
Indian Institute of Technology Kanpur, India

The "River Publishers Series in Computing and Information Science and Technology" covers research which ushers the 21st Century into an Internet and multimedia era. Networking suggests transportation of such multimedia contents among nodes in communication and/or computer networks, to facilitate the ultimate Internet.

Theory, technologies, protocols and standards, applications/services, practice and implementation of wired/wireless networking are all within the scope of this series. Based on network and communication science, we further extend the scope for 21st Century life through the knowledge in machine learning, embedded systems, cognitive science, pattern recognition, quantum/biological/molecular computation and information processing, user behaviors and interface, and applications across healthcare and society.

Books published in the series include research monographs, edited volumes, handbooks and textbooks. The books provide professionals, researchers, educators, and advanced students in the field with an invaluable insight into the latest research and developments.

Topics included in the series are as follows:-

- Artificial intelligence
- Cognitive Science and Brian Science
- Communication/Computer Networking Technologies and Applications
- Computation and Information Processing
- Computer Architectures
- Computer networks
- Computer Science
- Embedded Systems
- Evolutionary computation
- Information Modelling
- Information Theory
- Machine Intelligence
- Neural computing and machine learning
- Parallel and Distributed Systems
- Programming Languages
- Reconfigurable Computing
- Research Informatics
- Soft computing techniques
- Software Development
- Software Engineering
- Software Maintenance

For a list of other books in this series, visit www.riverpublishers.com

Transformative AI
Responsible, Transparent, and Trustworthy AI Systems

Ahmed Banafa

San Jose State University, CA, USA

Routledge
Taylor & Francis Group

NEW YORK AND LONDON

Published 2024 by River Publishers
River Publishers
Alsbjergvej 10, 9260 Gistrup, Denmark
www.riverpublishers.com

Distributed exclusively by Routledge
605 Third Avenue, New York, NY 10017, USA
4 Park Square, Milton Park, Abingdon, Oxon OX14 4RN

Transformative AI / by Ahmed Banafa.

Routledge is an imprint of the Taylor & Francis Group, an informa business

ISBN 978-87-7004-019-8 (hardback)
ISBN 978-87-7004-071-6 (paperback)
ISBN 978-10-0382-952-2 (online)
ISBN 978-10-3266-918-2 (master ebook)

While every effort is made to provide dependable information, the publisher, authors, and editors cannot be held responsible for any errors or omissions.

I dedicate this book to my amazing wife for all her love and help.

Contents

Preface

Artificial intelligence (AI) has become one of the most transformative technologies of the 21st century, with a potential to revolutionize various industries and reshape our lives in ways we could have never imagined. From chatbots and virtual assistants to self-driving cars and medical diagnoses, AI has proven to be an invaluable tool in enhancing productivity, efficiency, and accuracy across different domains.

This book is an attempt to provide a comprehensive overview of the field of artificial intelligence. It is designed to be an accessible resource for beginners, students, and professionals alike, who are interested in understanding the concepts, applications, and implications of AI.

The book is divided into two parts, each covering a distinct aspect of AI. The first part introduces the basic concepts of AI, including the history, types, components of AI, and the most common techniques and algorithms used in AI, such as machine learning, deep learning, natural language processing, and robotics. The second part delves into the various applications of AI in different fields, such as IoT, blockchain, quantum computing, robotics, and autonomous cars. In addition, the book covers the ethical and social implications of AI, such as bias, privacy, and job displacement.

This book aims to provide a balanced perspective on AI, presenting its opportunities as well as its challenges. It also includes real-world examples and case studies to help readers understand how AI is being used in practice.

I hope that this book will serve as a useful resource for readers looking to learn more about this exciting and rapidly evolving field.

Readership:
Technical and nontechnical audience including: C-level executives, directors, lawyers, journalists, sales and marketing professionals, engineers, developers, and students.

List of Figures

List of Abbreviations

ACI	Autonomic computing initiative
AGI	Artificial general intelligence
AI	Artificial intelligence
CAGR	Compound annual growth rate
CIS	Center for Internet Security
CNN	Convolutional neural network
DX	Digital transformation
EM	Electro magnetic
GAN	Generative adversarial network
GIGO	Garbage in garbage out
GPT	Generative pre-trained transformer
IoE	Internet of Everything
IoT	Internet of Things
LSTM	Long short-term memory
ML	Machine learning
MLP	Machine learning poisoning
NER	Named entity recognition
NLG	Natural language generation
NLP	Natural language processing
OCR	Optical character recognition
OOV	Out-of-vocabulary
POS	Part-of-speech
PSS	Privacy/security/safety
QCaaS	Quantum Computing as a Service
QML	Quantum machine learning
RNN	Recurrent neural network
SSP	Security, safety, privacy
TCO	Total cost of ownership
VAE	Variational autoencoder
WCD	Wearable computing devices

Part I

Understanding AI

1

What is AI?

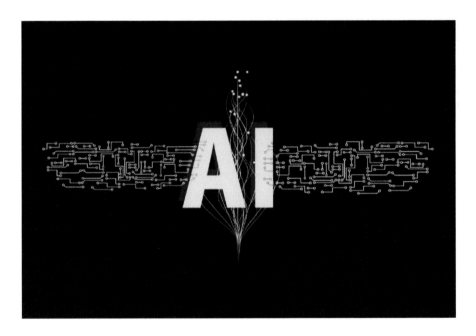

The field of artificial intelligence (AI) has a rich and fascinating history that stretches back over six decades. The story of AI is one of scientific inquiry, technological innovation, and the evolution of human thought about what it means to be intelligent. In this article, we will explore the major milestones in the history of AI and examine how this exciting field has evolved over time.

The origins of AI can be traced back to the 1950s, when researchers began to explore the possibility of creating machines that could think and reason like humans. One of the key figures in the early development of AI was the mathematician and logician Alan Turing. In his seminal paper "Computing

Machinery and Intelligence," Turing proposed a test that would measure a machine's ability to exhibit intelligent behavior that was indistinguishable from that of a human. This test, known as the Turing test, became a central concept in the study of AI.

In the years following Turing's work, researchers made significant progress in developing early AI systems. One of the earliest examples was the Logic Theorist, a program developed by Allen Newell and J.C. Shaw in 1955 that could prove mathematical theorems. Another landmark achievement was the creation of the General Problem Solver by Newell and Herbert Simon in 1957. This program was able to solve a range of problems by searching through a set of rules and making logical deductions.

The 1960s saw a surge of interest in AI, with researchers working to develop more advanced systems and applications. One notable example was the development of the first expert system, called Dendral, by Edward Feigenbaum and Joshua Lederberg in 1965. This system was designed to identify the molecular structure of organic compounds, and it proved to be highly successful in its field.

The 1970s and 1980s saw further advances in AI, with researchers developing new algorithms and techniques for machine learning, natural language processing, and computer vision. One of the key breakthroughs during this period was the development of the first neural network, which was inspired by the structure of the human brain. Another important development was the creation of rule-based expert systems, which allowed computers to make decisions based on a set of predefined rules.

In the 1990s and 2000s, AI continued to evolve rapidly, with researchers developing new techniques for deep learning, natural language processing, and image recognition. One of the most significant breakthroughs during this period was the development of the first autonomous vehicles, which used a combination of sensors, machine learning algorithms, and computer vision to navigate roads and highways.

Today, AI is a rapidly growing field that is transforming the way we live and work. From virtual assistants like Siri and Alexa to self-driving cars and sophisticated medical imaging systems, AI is playing an increasingly important role in our daily lives. Some of the most exciting developments in AI today include the development of advanced robots, the creation of AI-powered medical diagnosis tools, and the use of machine learning algorithms to analyze large amounts of data and identify patterns.

In conclusion, the history of AI is a story of scientific inquiry, technological innovation, and the evolution of human thought about what it means to be intelligent. From the early work of Turing and Newell to the breakthroughs of the present day, AI has come a long way in just a few short decades. As we look into the future, it is clear that AI will continue to play an increasingly important role in shaping the world around us, and we can only imagine what new breakthroughs and innovations lie ahead.

Artificial intelligence (AI) is a term that refers to the capability of machines or computer programs to perform tasks that typically require human intelligence, such as learning, problem-solving, and decision-making. The field of AI has been around for several decades, but recent advances in technology have led to a rapid increase in its capabilities and applications. In this essay, we will explore what AI is, how it works, and its current and potential future applications.

AI can be broadly categorized into two types: narrow or weak AI, and general or strong AI. Narrow AI is designed to perform specific tasks or solve specific problems, such as image recognition or language translation. It relies on machine learning algorithms that are trained on large datasets to recognize patterns and make predictions based on those patterns. General AI, on the other hand, aims to replicate human intelligence in a broad range of domains, such as reasoning, perception, and creativity. While narrow AI has made significant progress in recent years, general AI remains a long-term goal.

AI works by processing large amounts of data, identifying patterns and relationships, and using that information to make decisions or take actions. This process is made possible by algorithms that are designed to learn from the data they are fed. One of the most common types of AI algorithms is neural networks, which are modeled after the structure and function of the human brain. These networks consist of layers of interconnected nodes that process information and pass it on to the next layer, allowing the algorithm to identify patterns and relationships in the data.

AI has a wide range of applications across various industries and domains. In healthcare, for example, AI is being used to improve disease diagnosis and treatment by analyzing medical images and patient data. In finance, AI is being used to analyze market trends and make investment decisions. In transportation, AI is being used to optimize traffic flow and improve road safety. In education, AI is being used to personalize learning and improve student outcomes. These are just a few examples of the many ways in which AI is already being used to improve our lives.

Despite its many potential benefits, AI also raises several ethical and societal concerns. One of the biggest concerns is the potential impact of AI on employment. As AI becomes more advanced, it may be able to perform many tasks that are currently done by humans, leading to significant job losses. There are also concerns about the potential misuse of AI, such as the use of facial recognition technology for surveillance purposes. Additionally, there are concerns about the bias that may be built into AI algorithms, as they are only as unbiased as the data they are trained on.

Looking into the future, AI is likely to play an increasingly important role in our lives. As AI becomes more advanced, it will be able to perform more complex tasks and make more sophisticated decisions. This will have implications for virtually every aspect of society, from healthcare and finance to transportation and education. It will be important for policymakers and society as a whole to grapple with the ethical and societal implications of these developments and ensure that AI is used for the benefit of all.

In conclusion, AI is a rapidly advancing field that has the potential to revolutionize many aspects of our lives. It relies on algorithms that are designed to learn from data and identify patterns, and it has a wide range of applications across various industries and domains. While it raises many ethical and societal concerns, it also has the potential to bring significant benefits to the society. As AI continues to advance, it will be important for the society to grapple with these issues and ensure that its development is guided by ethical principles and used for the benefit of all.

What is Machine Learning?

Machine learning is a subset of artificial intelligence (AI) that involves the use of algorithms and statistical models to enable machines to learn from data and improve their performance on specific tasks. It is a rapidly growing field with numerous applications across various industries, including healthcare, finance, transportation, and more. In this essay, we will explore what machine learning is, how it works, and its current and potential future applications.

At its core, machine learning involves the use of algorithms that enable machines to learn from data. These algorithms are designed to identify patterns and relationships in the data and use that information to make predictions or take actions. One of the most common types of machine learning algorithms is supervised learning, which involves training a model on a labeled dataset, where the correct output is known for each input. The model then uses this training data to make predictions on new, unseen data.

Another type of machine learning is unsupervised learning, which involves training a model on an unlabeled dataset and allowing it to identify patterns and relationships on its own. This type of learning is often used in applications such as clustering, where the goal is to group similar items together.

Reinforcement learning is another type of machine learning, which involves training a model to make decisions based on feedback from its environment. In this type of learning, the model learns through trial and error, adjusting its actions based on the rewards or punishments it receives.

Machine learning has numerous applications across various industries. In healthcare, for example, machine learning is being used to analyze medical images and patient data to improve disease diagnosis and treatment. In finance, machine learning is being used to analyze market trends and make investment decisions. In transportation, machine learning is being used to optimize traffic flow and improve road safety. In education, machine learning is being used to personalize learning and improve student outcomes. These are just a few examples of the many ways in which machine learning is being used to improve our lives.

One of the key advantages of machine learning is its ability to improve over time. As the machine is exposed to more data, it is able to improve its performance on specific tasks. This is known as "learning by experience." Machine learning models can also be used to make predictions or decisions in real time, allowing them to be used in applications where speed is critical.

Despite its many benefits, machine learning also raises several ethical and societal concerns. One of the biggest concerns is the potential bias that may be built into machine learning algorithms. If the training data is biased, the model will also be biased, potentially leading to unfair decisions or outcomes. Another concern is the potential impact of machine learning on employment, as machines may be able to perform many tasks that are currently done by humans, leading to significant job losses.

Looking into the future, machine learning is likely to play an increasingly important role in our lives. As the amount of data being generated continues to increase, machine learning will be essential for making sense of this data and extracting valuable insights. It will also be important for the society to grapple with the ethical and societal implications of these developments and ensure that machine learning is used for the benefit of all.

In conclusion, machine learning is a rapidly growing field that involves the use of algorithms and statistical models to enable machines to learn from data and improve their performance on specific tasks. It has numerous applications across various industries and has the potential to revolutionize many aspects of our lives. While it raises many ethical and societal concerns, it also has the potential to bring significant benefits to society. As machine learning continues to advance, it will be important for the society to grapple with these issues and ensure that its development is guided by ethical principles and used for the benefit of all.

What is Deep Learning?

Deep learning is a subset of machine learning that involves the use of artificial neural networks to enable machines to learn from data and perform complex tasks. It is a rapidly growing field that has revolutionized many industries, including healthcare, finance, and transportation. In this essay, we will explore what deep learning is, how it works, and its current and potential future applications.

At its core, deep learning involves the use of artificial neural networks to enable machines to learn from data. These networks are inspired by the structure of the human brain and are composed of layers of interconnected nodes that process information. Each node in the network performs a simple calculation based on its inputs and outputs a result, which is then passed on to the next layer of nodes.

The process of training a deep learning model involves feeding it a large dataset and adjusting the weights of the connections between the nodes to minimize the difference between the predicted output and the actual output. This process is known as backpropagation and involves iteratively adjusting the weights of the connections between the nodes until the model is able to accurately predict the output for a given input.

One of the key advantages of deep learning is its ability to perform complex tasks, such as image recognition, speech recognition, and natural language processing. For example, deep learning models can be trained to recognize faces in images, transcribe speech to text, and generate human-like responses to natural language queries.

Deep learning has numerous applications across various industries. In healthcare, for example, deep learning is being used to analyze medical images and identify potential diseases. In finance, deep learning is being used to analyze market trends and make investment decisions. In transportation, deep learning is being used to develop self-driving cars. These are just a few examples of the many ways in which deep learning is being used to improve our lives.

One of the key challenges of deep learning is the amount of data required to train the models. Deep learning models typically require large amounts of labeled data to accurately learn the underlying patterns and relationships in the data. This can be a challenge in industries such as healthcare, where labeled data may be scarce or difficult to obtain.

Another challenge is the complexity of deep learning models. Deep learning models can be difficult to interpret, and it can be challenging to understand why they make the predictions or decisions that they do. This can be a concern in applications where the decisions made by the model may have significant consequences.

Looking into the future, deep learning is likely to play an increasingly important role in our lives. As the amount of data being generated continues to increase, deep learning will be essential for making sense of this data and extracting valuable insights. It will also be important for the society to grapple with the ethical and societal implications of these developments and ensure that deep learning is used for the benefit of all.

In conclusion, deep learning is a rapidly growing field that involves the use of artificial neural networks to enable machines to learn from data and perform complex tasks. It has numerous applications across various industries and has the potential to revolutionize many aspects of our lives. While it raises many ethical and societal concerns, it also has the potential to bring significant benefits to the society. As deep learning continues to advance, it will be important for society to grapple with these issues and ensure that its development is guided by ethical principles and used for the benefit of all.

2

Neural Networks

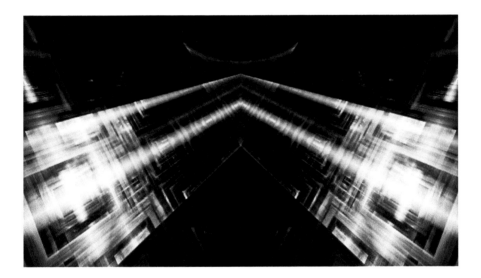

Neural networks are a type of artificial intelligence that has been inspired by the structure and function of the human brain. They are a computational model that is capable of learning from data and making predictions or decisions based on that learning. Neural networks are used in a variety of applications such as image recognition, speech recognition, natural language processing, and many more.

What is a Neural Network?

At its most basic level, a neural network is a collection of interconnected nodes, called neurons, which work together to process and analyze data. Each neuron in a neural network receives inputs from other neurons and produces an output that is transmitted to other neurons in the network. The inputs to

11

each neuron are weighted, meaning that some inputs are more important than others in determining the output of the neuron.

Neurons in a neural network are organized into layers, with each layer having a specific function in processing the data. The input layer is where the data is initially introduced into the network. The output layer produces the final result of the network's processing. Between the input and output layers are one or more hidden layers, which perform the majority of the processing in the network.

Training a Neural Network

The ability of a neural network to learn from data is what makes it so powerful. In order to train a neural network, a set of training data is provided to the network, along with the desired outputs for that data. The network then adjusts the weights of the connections between neurons to minimize the difference between the predicted output and the desired output. This process is known as backpropagation.

The number of layers and the number of neurons in each layer are important factors in determining the accuracy and speed of a neural network. Too few neurons or layers may result in the network being unable to accurately represent the complexity of the data being processed. Too many neurons or layers may result in overfitting, where the network is too specialized to the training data and unable to generalize to new data.

Applications of Neural Networks

Neural networks have a wide range of applications in areas such as image recognition, speech recognition, natural language processing, and many more. In image recognition, for example, a neural network can be trained to identify specific features of an image, such as edges or shapes, and use that information to recognize objects in the image.

In speech recognition, a neural network can be trained to identify the individual phonemes that make up words and use that information to transcribe spoken words into text. In natural language processing, a neural network can be trained to understand the meaning of words and sentences and use that understanding to perform tasks such as language translation or sentiment analysis.

Neural networks are a powerful tool for artificial intelligence, which are capable of learning from data and making predictions or decisions based on

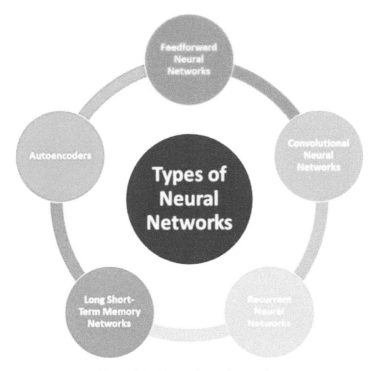

Figure 2.1   Types of neural networks.

that learning. They are modeled after the structure and function of the human brain and have a wide range of applications in areas such as image recognition, speech recognition, and natural language processing. With continued research and development, neural networks have the potential to revolutionize the way we interact with technology and each other.

Types of Neural Networks

Neural networks are a subset of machine learning algorithms that are inspired by the structure and function of the human brain. They are capable of learning from data and making predictions or decisions based on that learning. There are several different types of neural networks, each with their own unique characteristics and applications (Figure 2.1).

1. **Feedforward neural networks:**
   Feedforward neural networks are the most basic type of neural network. They consist of one or more layers of neurons that are connected in a

feedforward manner, meaning that the output of each neuron is passed as input to the next layer of neurons. Feedforward neural networks are typically used for classification or regression tasks, where the output is a single value or class label.

2. **Convolutional neural networks:**
Convolutional neural networks (CNNs) are a type of neural network that is commonly used in image and video recognition tasks. They consist of several layers of neurons that are designed to process images or videos in a hierarchical manner. The first layer of a CNN is typically a convolutional layer, which applies a series of filters to the input image to extract features such as edges or textures. The output of the convolutional layer is then passed to one or more fully connected layers for further processing and classification.

3. **Recurrent neural networks:**
Recurrent neural networks (RNNs) are a type of neural network that is designed to handle sequential data, such as speech or text. They consist of a series of interconnected neurons, with each neuron in the network receiving inputs from both the previous neuron in the sequence and the current input. RNNs are capable of learning long-term dependencies in the input data, making them well-suited for tasks such as language translation or speech recognition.

4. **Long short-term memory networks:**
Long short-term memory (LSTM) networks are a type of RNN that is specifically designed to handle long-term dependencies in sequential data. They consist of a series of memory cells that can be selectively read, written to, or erased based on the input data. LSTMs are well-suited for tasks such as speech recognition or handwriting recognition, where the input data may have long-term dependencies that need to be captured.

5. **Autoencoders:**
Autoencoders are a type of neural network that is used for unsupervised learning tasks, such as feature extraction or data compression. They consist of an encoder network that takes the input data and maps it to a lower-dimensional latent space, and a decoder network that reconstructs the original input data from the latent space. Autoencoders are often used for tasks such as image or text generation, where the output is generated based on the learned features of the input data.

Neural networks are a powerful tool for machine learning that are capable of learning from data and making predictions or decisions based on that learning. There are several different types of neural networks, each with their own unique characteristics and applications. Feedforward neural networks are the most basic type of neural network, while convolutional neural networks are commonly used in image and video recognition tasks. Recurrent neural networks and long short-term memory networks are designed to handle sequential data, while autoencoders are used for unsupervised learning tasks such as feature extraction or data compression. With continued research and development, neural networks have the potential to revolutionize the way we interact with technology and each other.

3

Natural Language Processing (NLP)

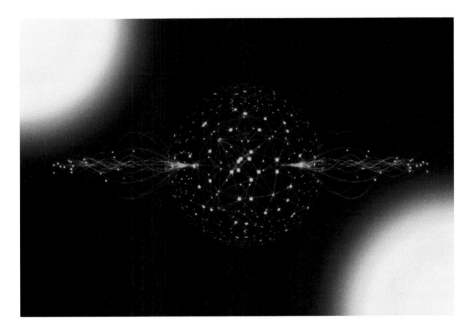

Natural language processing (NLP) is a subfield of artificial intelligence (AI) that focuses on the interaction between computers and humans using natural language. It involves analyzing, understanding, and generating human language in a way that is meaningful and useful to humans.

NLP has made significant strides in recent years, with the development of deep learning algorithms and the availability of large amounts of data. These advances have enabled NLP to be used in a wide range of applications, from language translation to chatbots to sentiment analysis.

In this article, we will explain the basics of NLP, its applications, and some of the challenges that researchers face.

The Basics of NLP

NLP involves several tasks, including:

1. Text classification: Assigning a category to a given text (e.g., spam or not spam).
2. Named entity recognition (NER): Identifying and classifying entities in a text, such as people, organizations, and locations.
3. Sentiment analysis: Determining the sentiment expressed in a text, whether it is positive, negative, or neutral.
4. Machine translation: Translating text from one language to another.
5. Question answering: Answering questions posed in natural language.
6. Text summarization: Generating a shorter summary of a longer text.

To accomplish these tasks, NLP algorithms use various techniques, such as:

1. Tokenization: Breaking a text into smaller units (tokens), such as words or subwords.
2. Part-of-speech (POS) tagging: Assigning a part of speech (e.g., noun, verb, adjective) to each token in a text.
3. Dependency parsing: Identifying the grammatical relationships between words in a sentence.
4. Named entity recognition: Identifying and classifying named entities in a text.
5. Sentiment analysis: Analyzing the tone of a text to determine whether it is positive, negative, or neutral.

Applications of NLP

NLP has numerous applications, including:

1. Chatbots: NLP can be used to build chatbots that can understand and respond to natural language queries.
2. Sentiment analysis: NLP can be used to analyze social media posts, customer reviews, and other texts to determine the sentiment expressed.
3. Machine translation: NLP can be used to translate text from one language to another, enabling communication across language barriers.
4. Voice assistants: NLP can be used to build voice assistants, such as Siri or Alexa, that can understand and respond to voice commands.
5. Text summarization: NLP can be used to generate summaries of longer texts, such as news articles or research papers.

Challenges in NLP

Despite the progress made in NLP, several challenges remain. Some of these challenges include:

1. Data bias: NLP algorithms can be biased if they are trained on data that is not representative of the population.
2. Ambiguity: Natural language can be ambiguous, and NLP algorithms must be able to disambiguate text based on context.
3. Out-of-vocabulary (OOV) words: NLP algorithms may struggle with words that are not in their training data.
4. Language complexity: Some languages, such as Chinese and Arabic, have complex grammatical structures that can make NLP more challenging.
5. Understanding the context: NLP algorithms must be able to understand the context in which a text is written to correctly interpret its meaning.

NLP is a rapidly advancing field that has numerous applications, from chatbots to machine translation to sentiment analysis. As NLP continues to improve, we can expect to see more sophisticated language-based technologies emerge, such as virtual assistants that can understand and respond to complex queries. However, researchers must continue to address the challenges associated with NLP.

Natural Language Generation (NLG)

Natural language generation (NLG) is a subfield of natural language processing (NLP) that focuses on the automatic generation of human-like text. It involves transforming structured data into natural language that is understandable to humans. NLG has numerous applications, including automated report writing, chatbots, and personalized messaging.

In this article, we will explain the basics of NLG, its applications, and some of the challenges that researchers face.

The Basics of NLG

NLG involves several tasks, including:

1. Content determination: Deciding what information to include in the generated text.
2. Text structuring: Organizing the content into a coherent structure, such as paragraphs or bullet points.

3. Lexicalization: Selecting appropriate words and phrases to convey the intended meaning.
4. Referring expression generation: Determining how to refer to entities mentioned in the text, such as using pronouns or full names.
5. Sentence planning: Deciding how to structure individual sentences, such as choosing between active and passive voice.
6. Realization: Generating the final text.

To accomplish these tasks, NLG algorithms use various techniques, such as:

1. Template-based generation: Using pre-defined templates to generate text.
2. Rule-based generation: Applying rules to generate text.
3. Statistical-based generation: Using statistical models to generate text based on patterns in the data.
4. Neural-based generation: Using deep learning models to generate text.

Applications of NLG

NLG has numerous applications, including:

1. Automated report writing: NLG can be used to automatically generate reports, such as weather reports or financial reports, based on structured data.
2. Chatbots: NLG can be used to generate responses to user queries in natural language, enabling more human-like interactions with chatbots.
3. Personalized messaging: NLG can be used to generate personalized messages, such as marketing messages or product recommendations.
4. E-commerce: NLG can be used to automatically generate product descriptions based on structured data, improving the efficiency of e-commerce operations.
5. Content creation: NLG can be used to generate news articles or summaries based on structured data.

Challenges in NLG

Despite the progress made in NLG, several challenges remain. Some of these challenges include:

1. Data quality: NLG algorithms rely on high-quality structured data to generate accurate and useful text.
2. Naturalness: NLG algorithms must generate text that is natural-sounding and understandable to humans.
3. Domain specificity: NLG algorithms must be tailored to specific domains to generate accurate and useful text.
4. Personalization: NLG algorithms must be able to generate personalized text that is tailored to individual users.
5. Evaluation: Evaluating the quality of generated text is challenging, as it requires subjective judgments by humans.

NLG is a rapidly advancing field that has numerous applications, from automated report writing to chatbots to personalized messaging. As NLG continues to improve, we can expect to see more sophisticated and natural-sounding language-based technologies emerge, enabling more human-like interactions with machines. However, researchers must continue to address the challenges associated with NLG to ensure that generated text is accurate, natural-sounding, and useful.

4

Computer Vision

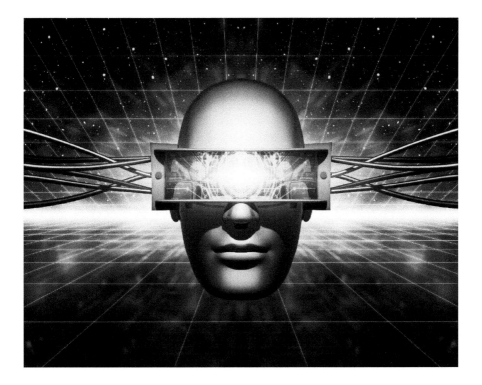

Computer vision is a field of study that enables computers to *interpret and understand* visual data from the world around us. This field has seen significant advancements in recent years, thanks to the increasing availability of *large datasets*, *powerful computing resources*, and *advanced algorithms*. In this article, we will explore the challenges and opportunities in the field of computer vision and its future.

What is Computer Vision?

Computer vision is a field of study that focuses on enabling computers to understand and interpret visual data from the world around us. It involves developing algorithms and techniques to enable computers to recognize, analyze, and process visual information from various sources such as images, videos, and depth maps.

The goal of computer vision is to enable machines *to see and understand the world in the same way humans do*. This requires the use of advanced algorithms and techniques that can extract useful information from visual data and use it to make decisions and take actions.

Applications of Computer Vision

Computer vision has a wide range of applications in various fields, some of which are listed below:

1. **Self-driving cars:** Computer vision is a key technology in the development of self-driving cars. It enables cars to detect and recognize objects such as pedestrians, other vehicles, and road signs, and make decisions based on that information.
2. **Facial recognition:** Computer vision is also used in facial recognition systems, which are used for security and surveillance purposes. It enables cameras to detect and recognize faces, and match them to a database of known individuals.
3. **Medical imaging:** Computer vision is used in medical imaging to analyze and interpret medical images such as X-rays, CT scans, and MRI scans. It enables doctors to detect and diagnose diseases and injuries more accurately and quickly.
4. **Industrial automation:** Computer vision is used in industrial automation to monitor and control manufacturing processes. It enables machines to detect and identify parts and products, and perform quality control checks.

Challenges in Computer Vision

Despite the significant progress in computer vision, there are still many challenges that need to be overcome. One of the biggest challenges is *developing algorithms and techniques that can work in real-world environments* with a high degree of variability and uncertainty. For example, recognizing objects in images and videos can be challenging when the lighting conditions, camera angle, and object orientation vary.

Another challenge *is developing algorithms that can process and analyze large volumes of visual data in real time.* This is especially important in applications such as autonomous vehicles and robotics, where decisions need to be made quickly and accurately.

A third challenge *is developing algorithms that are robust to adversarial attacks.* Adversarial attacks are a type of attack where an attacker intentionally manipulates an image or video to deceive a computer vision system. For example, an attacker can add imperceptible noise to an image that can cause a computer vision system to misclassify an object.

Opportunities in Computer Vision

Despite the challenges, there are many opportunities in computer vision. One of the biggest opportunities is the *ability to automate tasks* that were previously performed by humans. For example, computer vision can be used to automate quality control in manufacturing, reduce errors in medical imaging, and improve safety in autonomous vehicles.

Another opportunity is *the ability to analyze large volumes of visual data* to gain insights and make predictions. For example, computer vision can be used to analyze satellite images to monitor crop health, detect changes in land use, and monitor environmental conditions.

A third opportunity *is the ability to develop new products and services* that can improve people's lives. For example, computer vision can be used to develop assistive technologies for the visually impaired, improve security and surveillance systems, and enhance virtual and augmented reality experiences.

Future of Computer Vision

The future of computer vision is very promising, and we can expect to see many new applications and advancements in this field. Some of the areas where we can expect to see significant progress include:

1. **Deep learning:** Deep learning is a subfield of machine learning that has shown remarkable progress in computer vision. We can expect to see continued advancements in deep learning algorithms, which will enable computers to recognize and interpret visual data more accurately and efficiently.

2. **Augmented and virtual reality:** Augmented and virtual realities are two areas where computer vision can have a significant impact. We can expect to see new applications and advancements in these areas that will enhance our experiences and improve our ability to interact with the world around us.

3. **Autonomous vehicles:** Autonomous vehicles are one of the most promising applications of computer vision. We can expect to see continued advancements in autonomous vehicle technology, which will enable safer and more efficient transportation.

Computer vision is a rapidly growing field that has many challenges and opportunities. Despite the challenges, we can expect to see many new applications and advancements in this field that will improve our ability to interpret and understand visual data from the world around us. As computer vision continues to evolve, we can expect to see new products and services that will improve people's lives and transform industries.

Future of Computer Vision and AI

The future of computer vision and AI is very promising, and we can expect to see many new applications and advancements in these fields. Some of the areas where we can expect to see significant progress include the following (Figure 4.1):

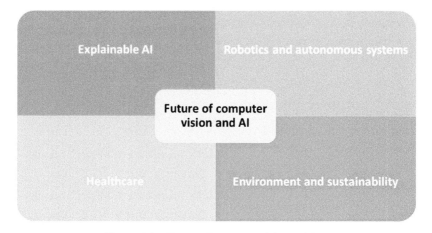

Figure 4.1   Future of computer vision and AI.

1. **Explainable AI:** Developing AI systems that can explain their reasoning and decision-making processes is critical for ensuring trust and transparency in AI systems. We can expect to see continued advancements in explainable AI, which will enable us to better understand and trust the decisions made by AI systems.
2. **Robotics and autonomous systems:** Robotics and autonomous systems are two areas where computer vision and AI can have a significant impact. We can expect to see continued advancements in these areas that will enable us to develop more advanced and capable robots and autonomous systems.
3. **Healthcare:** Healthcare is an area where computer vision and AI can be used to improve patient outcomes and reduce costs. We can expect to see continued advancements in medical imaging and diagnostics, personalized medicine, and drug discovery, which will help to improve patient care and reduce healthcare costs.
4. **Environment and sustainability:** Computer vision and AI can be used to monitor and manage the environment, including monitoring pollution, tracking wildlife populations, and monitoring climate change. We can expect to see continued advancements in these areas, which will help to improve our understanding of the environment and enable us to make better decisions about how to manage it.

Computer vision and AI are two rapidly growing fields that have many challenges and opportunities. Despite the challenges, we can expect to see many new applications and advancements in these fields that will improve our ability to understand and interpret data from the world around us. As computer vision and AI continue to evolve, we can expect to see new products and services that will improve people's lives and transform industries.

5

Levels of AI

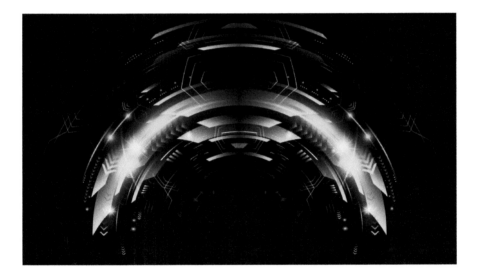

Artificial intelligence (AI) is one of the most rapidly advancing fields of technology today, with the potential to revolutionize virtually every industry and aspect of our daily lives. AI is often categorized into different levels based on their capabilities and autonomy, which can help us understand the current state of AI development and the challenges ahead (Figure 5.1).

Level 1: Reactive Machines: The simplest form of AI is the reactive machine, which only reacts to inputs without any memory or ability to learn from experience. Reactive machines are programmed to perform specific tasks and are designed to respond to particular situations in pre-defined ways. They do not have any past data or memory to draw from, and they do not have the ability to consider the wider context of their actions.

29

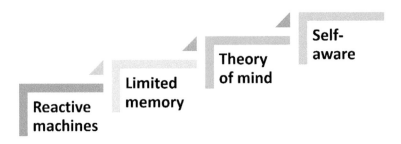

Figure 5.1 AI levels.

One example of a reactive machine is IBM's Deep Blue, which defeated the chess champion Garry Kasparov in 1997. Deep Blue used an algorithm to evaluate millions of possible moves and choose the best one based on a set of pre-defined rules. Another example is the Roomba robotic vacuum cleaner, which senses its environment and navigates around it but does not have the ability to remember its previous cleaning routes.

Level 2: Limited Memory: Limited memory AI can store past data and experiences to make informed decisions based on patterns and past experiences. This type of AI is commonly used in recommendation systems in e-commerce, where past purchases or browsing behavior is used to recommend future purchases.

Limited memory AI systems are designed to learn from experience and improve over time. For example, voice assistants such as Siri, Alexa, and Google Assistant use natural language processing and machine learning to understand and respond to user queries. These systems can also learn from past interactions with users, adapting to their preferences and improving their responses over time.

Level 3: Theory of Mind: Theory of Mind AI goes beyond reactive and limited memory systems, simulating human-like thoughts and emotions and having a deeper understanding of human behavior and social interactions. This type of AI is still in the research phase and has not yet been fully developed.

Theory of mind AI has the potential to revolutionize fields such as psychology and social sciences, by providing insights into how humans think and interact with each other. One example of theory of mind AI is the work being done by the MIT Media Lab's Affective Computing group, which is developing algorithms to recognize and respond to human emotions based on facial expressions, tone of voice, and other cues.

Level 4: Self-Aware: Self-aware AI is the most advanced level of AI, possessing the ability to not only understand human emotions and behaviors but also to be aware of their own existence and consciousness. This level of AI is still a theoretical concept and has not yet been achieved.

Self-aware AI has the potential to be truly transformative, with the ability to reflect on their own experiences and make autonomous decisions based on their own motivations and goals. However, the development of self-aware AI also raises important ethical and philosophical questions about the nature of consciousness, free will, and the relationship between humans and machines.

Conclusion

The levels of AI provide a useful framework for understanding the current state of AI development and the challenges ahead. While reactive and limited memory AI are already being used in many applications today, theory of mind and self-aware AI are still in the research phase and may take decades or even centuries to fully develop.

As AI continues to advance, it is important to consider the ethical, social, and philosophical implications of this technology. AI has the potential to transform our society and economy, but it also raises important questions about the nature of intelligence, consciousness, and human—machine interactions. By considering the levels of AI and their implications, we can better understand the opportunities and challenges ahead as we continue to push the boundaries of artificial intelligence.

6

Generative AI and Other Types of AI

AI refers to the development of computer systems that can perform tasks that typically require human intelligence, such as learning, reasoning, problem-solving, perception, and natural language understanding.

AI is based on the idea of creating intelligent machines that can work and learn like humans. These machines can be trained to recognize patterns, understand speech, interpret data, and make decisions based on that data. AI has many practical applications, including speech recognition, image recognition, natural language processing, autonomous vehicles, and robotics, to name a few.

Types of AI

Narrow AI, also known as weak AI, is an AI system designed to perform a specific task or set of tasks. These tasks are often well-defined and narrow in scope, such as image recognition, speech recognition, or language translation. Narrow AI systems rely on specific algorithms and techniques to solve problems and make decisions within their domain of expertise. These systems do not possess true intelligence, but rather mimic intelligent behavior within a specific domain.

General AI, also known as strong AI or human-level AI, is an AI system that can perform any intellectual task that a human can do. General AI would have the ability to reason, learn, and understand any intellectual task that a human can perform. It would be capable of solving problems in a variety of domains and would be able to apply its knowledge to new and unfamiliar situations. General AI is often thought of as the ultimate goal of AI research but is currently only a theoretical concept.

Super AI, also known as artificial superintelligence, is an AI system that surpasses human intelligence in all areas. Super AI would be capable of performing any intellectual task with ease and would have an intelligence level far beyond that of any human being. Super AI is often portrayed in science fiction as a threat to humanity, as it could potentially have its own goals and motivations that could conflict with those of humans. Super AI is currently only a theoretical concept, and the development of such a system is seen as a long-term goal of AI research.

Technical Types of AI (Figure 6.1)

1. **Rule-based AI:** Rule-based AI, also known as expert systems, is a type of AI that relies on a set of pre-defined rules to make decisions or recommendations. These rules are typically created by human experts in a particular domain and are encoded into a computer program. Rule-based AI is useful for tasks that require a lot of domain-specific knowledge, such as medical diagnosis or legal analysis.
2. **Supervised learning:** Supervised learning is a type of machine learning that involves training a model on a labeled dataset. This means that the dataset includes both input data and the correct output for each example. The model learns to map input data to output data and can then make

Figure 6.1   Technical types of AI.

predictions on new, unseen data. Supervised learning is useful for tasks such as image recognition or natural language processing.

3. **Unsupervised learning:** Unsupervised learning is a type of machine learning that involves training a model on an unlabeled dataset. This means that the dataset only includes input data, and the model must find patterns or structure in the data on its own. Unsupervised learning is useful for tasks such as clustering or anomaly detection.

4. **Reinforcement learning:** Reinforcement learning is a type of machine learning that involves training a model to make decisions based on rewards and punishments. The model learns by receiving feedback in the form of rewards or punishments based on its actions and adjusts its behavior to maximize its reward. Reinforcement learning is useful for tasks such as game playing or robotics.

5. **Deep learning:** Deep learning is a type of machine learning that involves training deep neural networks on large datasets. Deep neural networks are neural networks with multiple layers, allowing them to learn complex patterns and structures in the data. Deep learning is useful for tasks such as image recognition, speech recognition, and natural language processing.

6. **Generative AI:** Generative AI is a type of AI that is used to generate new content, such as images, videos, or text. It works by using a model that has been trained on a large dataset of examples and then uses this knowledge to generate new content that is similar to the examples it has been trained on. Generative AI is useful for tasks such as computer graphics, natural language generation, and music composition.

Generative AI

Generative AI is a type of artificial intelligence that is used to generate new content, such as images, videos, or even text. It works by using a model that has been trained on a large dataset of examples and then uses this knowledge to generate new content that is similar to the examples it has been trained on.

One of the most exciting applications of generative AI is in the field of computer graphics. By using generative models, it is possible to create realistic images and videos that look like they were captured in the real world. This can be incredibly useful for a wide range of applications, from creating realistic game environments to generating lifelike product images for e-commerce websites.

Another application of generative AI is in the field of natural language processing. By using generative models, it is possible to generate new text that is similar in style and tone to a particular author or genre. This can be useful for a wide range of applications, from generating news articles to creating marketing copy.

One of the key advantages of generative AI is its ability to create new content that is both creative and unique. Unlike traditional computer programs, which are limited to following a fixed set of rules, generative AI is able to learn from examples and generate new content that is similar, but not identical, to what it has seen before. This can be incredibly useful for applications where creativity and originality are important, such as in the arts or in marketing.

However, there are also some potential drawbacks to generative AI. One of the biggest challenges is ensuring that the content generated by these models is not biased or offensive. Because these models are trained on a dataset of examples, they may inadvertently learn biases or stereotypes that are present in the data. This can be especially problematic in applications like natural language processing, where biased language could have real-world consequences.

Another challenge is ensuring that the content generated by these models is of high quality. Because these models are based on statistical patterns in the data, they may occasionally produce outputs that are nonsensical or even offensive. This can be especially problematic in applications like chatbots or customer service systems, where errors or inappropriate responses could damage the reputation of the company or organization.

Despite these challenges, however, the potential benefits of generative AI are enormous. By using generative models, it is possible to create new content that is both creative and unique, while also being more efficient and cost-effective than traditional methods. With continued research and development, generative AI could play an increasingly important role in a wide range of applications, from entertainment and marketing to scientific research and engineering.

One of the challenges in creating effective generative AI models is choosing the right architecture and training approach. There are many different types of generative models, each with its own strengths and weaknesses. Some of the most common types of generative models include variational autoencoders, generative adversarial networks, and autoregressive models.

Variational autoencoders are a type of generative model that uses an encoder−decoder architecture to learn a compressed representation of the input data, which can then be used to generate new content. This approach is useful for applications where the input data is high-dimensional, such as images or video.

Generative adversarial networks (GANs) are another popular approach to generative AI. GANs use a pair of neural networks to generate new content. One network generates new content, while the other network tries to distinguish between real and fake content. By training these networks together, GANs are able to generate content that is both realistic and unique.

Autoregressive models are a type of generative model that uses a probabilistic model to generate new content. These models work by predicting the probability of each output.

Risks of Generative AI

Generative AI has the potential to revolutionize many industries and bring about numerous benefits, but it also poses several risks that need to be addressed. Following are some of the most significant risks associated with generative AI (Figure 6.2):

1. **Misuse:** Generative AI can be used to create fake content that is difficult to distinguish from real content, such as deepfakes. This can be used to spread false information or create misleading content that could have serious consequences.
2. **Bias:** Generative AI systems learn from data, and if the data is biased, then the system can also be biased. This can lead to unfair or discriminatory outcomes in areas such as hiring or lending.
3. **Security:** Generative AI can be used to create new forms of cyberattacks, such as creating realistic phishing emails or malware that can evade traditional security measures.
4. **Intellectual property:** Generative AI can be used to create new works that may infringe on the intellectual property of others, such as using existing images or music to generate new content.
5. **Privacy:** Generative AI can be used to create personal information about individuals, such as realistic images or videos that could be used for identity theft or blackmail.
6. **Unintended consequences:** Generative AI systems can create unexpected or unintended outcomes, such as creating new types of malware or causing harm to individuals or society.

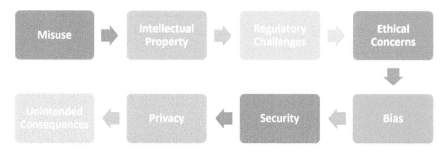

Figure 6.2   Risks of generative AI.

7. **Regulatory challenges:** The use of generative AI raises regulatory challenges related to its development, deployment, and use, including questions around accountability and responsibility.
8. **Ethical concerns:** The use of generative AI raises ethical concerns, such as whether it is appropriate to create content that is indistinguishable from real content or whether it is ethical to use generative AI for military or surveillance purposes.

Future of Generative AI

Generative AI is a rapidly advancing field that holds enormous potential for many different applications. As the technology continues to develop, we can expect to see some exciting advancements and trends in the future of generative AI. Following are some possible directions for the field. *Improved Natural Language Processing* (NLP): Natural language processing is one area where generative AI is already making a big impact, and we can expect to see this trend continue in the future. Advancements in NLP will allow for more natural-sounding and contextually appropriate responses from chatbots, virtual assistants, and other AI-powered communication tools. *Increased Personalization*: As generative AI systems become more sophisticated, they will be able to generate content that is more tailored to individual users. This could mean everything from personalized news articles to custom video game levels that are generated on the fly. *Enhanced Creativity*: Generative AI is already being used to generate music, art, and other forms of creative content. As the technology improves, we can expect to see more and more AI-generated works of art that are indistinguishable from those created by humans. *Better Data Synthesis*: As datasets become increasingly complex, generative AI will become an even more valuable tool for synthesizing and generating new data. This could be especially important in scientific research, where AI-generated data could help researchers identify patterns and connections that might otherwise go unnoticed. *Increased Collaboration*: One of the most exciting possibilities for generative AI is its potential to enhance human creativity and collaboration. By providing new and unexpected insights, generative AI could help artists, scientists, and other creatives work together in novel ways to generate new ideas and solve complex problems.

The future of generative AI looks bright, with plenty of opportunities for innovation and growth in the years ahead.

7

Generative AI: Types, Skills, Opportunities, and Challenges

Generative AI refers to a class of machine learning techniques that aim to generate new data that is similar to, but not identical to, the data it was trained on. In other words, generative AI models learn to create new data samples that have similar statistical properties to the training data, allowing them to create new content such as images, videos, audio, or text that has never been seen before.

There are several types of generative AI models, including the following (Figure 7.1):

1. **Variational autoencoders (VAEs):** A VAE is a type of generative model that learns to encode input data into a lower-dimensional latent space and then decode the latent space back into an output space to generate new

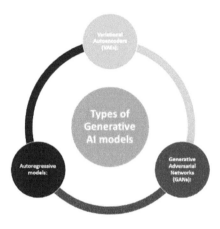

Figure 7.1   Types of generative AI models.

data that is similar to the original input data. VAEs are commonly used for image and video generation.

2. **Generative adversarial networks (GANs):** A GAN is a type of generative model that learns to generate new data by pitting two neural networks against each other − a generator and a discriminator. The generator learns to create new data samples that can fool the discriminator, while the discriminator learns to distinguish between real and fake data samples. GANs are commonly used for image, video, and audio generation.

3. **Autoregressive models:** Autoregressive models are a type of generative model that learns to generate new data by predicting the probability distribution of the next data point given the previous data points. These models are commonly used for text generation.

Skills Needed to Work in Generative AI

1. **Strong mathematical and programming skills:** In generative AI, you will be working with complex algorithms and models that require a solid understanding of mathematical concepts such as linear algebra, calculus, probability theory, and optimization algorithms. Additionally, you will need to be proficient in programming languages such as Python,

TensorFlow, PyTorch, or Keras, which are commonly used in generative AI research and development.

2. **Deep learning expertise:** Generative AI involves the use of deep learning techniques and frameworks, which require a deep understanding of how they work. You should have experience with various deep learning models, such as convolutional neural networks (CNNs), recurrent neural networks (RNNs), and transformer-based models, as well as experience with training, fine-tuning, and evaluating these models.

3. **Understanding of natural language processing (NLP):** If you are interested in generative AI for NLP, you should have experience with NLP techniques such as language modeling, text classification, sentiment analysis, and machine translation. You should also be familiar with NLP-specific deep learning models, such as transformers and encoder–decoder models.

4. **Creative thinking:** In generative AI, you will be tasked with generating new content, such as images, music, or text. This requires the ability to think creatively and come up with innovative ideas for generating content that is both novel and useful.

5. **Data analysis skills:** Generative AI requires working with large datasets; so you should have experience with data analysis and visualization techniques. You should also have experience with data preprocessing, feature engineering, and data augmentation to prepare data for training and testing models.

6. **Collaboration skills:** Working in generative AI often requires collaborating with other team members, such as data scientists, machine learning engineers, and designers. You should be comfortable working in a team environment and communicating technical concepts to non-technical stakeholders.

7. **Strong communication skills:** As a generative AI expert, you will be communicating complex technical concepts to both technical and non-technical stakeholders. You should have strong written and verbal communication skills to effectively explain your work and findings to others.

8. **Continuous learning:** Generative AI is a rapidly evolving field, and staying up-to-date with the latest research and techniques is essential to stay competitive. You should have a strong appetite for continuous learning and be willing to attend conferences, read research papers, and experiment with new techniques to improve your skills.

Overall, working in generative AI requires a mix of technical, creative, and collaborative skills. By developing these skills, you will be well-equipped to tackle challenging problems in this exciting and rapidly evolving field.

Generative AI Opportunities (Figure 7.2)

1. **Creative content generation:** One of the most exciting opportunities in generative AI is the ability to create new and unique content in various domains such as art, music, literature, and design. Generative AI can

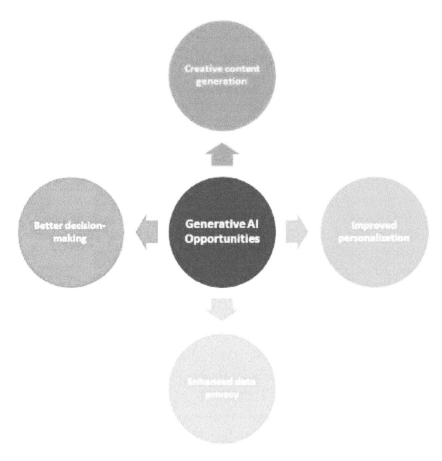

Figure 7.2   Generative AI opportunities.

help artists and designers to create new and unique pieces of work that may not have been possible otherwise.

2. **Improved personalization:** Generative AI can also help businesses to provide more personalized experiences to their customers. For example, it can be used to generate personalized recommendations, product designs, or content for users based on their preferences.

3. **Enhanced data privacy:** Generative AI can be used to generate synthetic data that mimics the statistical properties of real data, which can be used to protect users' privacy. This can be particularly useful in healthcare, where sensitive medical data needs to be protected.

4. **Better decision-making:** Generative AI can also be used to generate alternative scenarios to help decision-makers make better-informed decisions. For example, it can be used to simulate different scenarios in finance, weather forecasting, or traffic management.

Generative AI Challenges (Figure 7.3)

1. **Data quality:** Generative AI models heavily rely on the quality and quantity of data used to train them. Poor-quality data can result in models

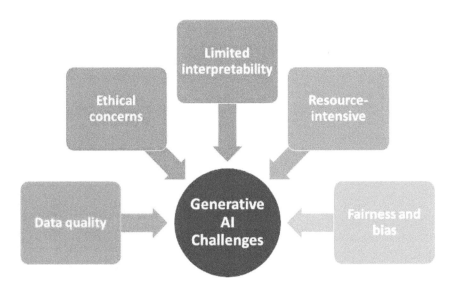

Figure 7.3 Generative AI challenges.

that generate low-quality outputs, which can impact their usability and effectiveness.

2. **Ethical concerns:** Generative AI can raise ethical concerns around the use of synthesized data, particularly in areas such as healthcare, where synthetic data may not accurately reflect real-world data. Additionally, generative AI can be used to create fake media, which can have negative consequences if misused.

3. **Limited interpretability:** Generative AI models can be complex and difficult to interpret, making it hard to understand how they generate their outputs. This can make it difficult to diagnose and fix errors or biases in the models.

4. **Resource-intensive:** Generative AI models require significant computing power and time to train, making it challenging to scale them for large datasets or real-time applications.

5. **Fairness and bias:** Generative AI models can perpetuate biases present in the training data, resulting in outputs that are discriminatory or unfair to certain groups. Ensuring fairness and mitigating biases in generative AI models is an ongoing challenge.

Generative AI has numerous applications in various fields, including art, design, music, and literature. For example, generative AI models can be used to create new art, design new products, compose new music, or write new stories. Generative AI is also used in healthcare for generating synthetic medical data to protect patients' privacy, or in cybersecurity to generate fake data to test security systems.

ChatGPT

ChatGPT is a specific implementation of generative AI. Generative AI is a broad category of machine learning algorithms that are used to generate new data based on patterns learned from existing data. In the context of natural language processing, generative AI can be used to generate new text based on patterns learned from a large corpus of text data.

ChatGPT is a specific implementation of generative AI that is designed to generate text in response to user input in a conversational setting. ChatGPT is based on the GPT (generative pre-trained transformer) architecture, which is a type of neural network that has been pre-trained on a massive amount of text data. This pre-training allows ChatGPT to generate high-quality text that is both fluent and coherent.

In other words, ChatGPT is a specific application of generative AI that is designed for conversational interactions. Other applications of generative AI may include language translation, text summarization, or content generation for marketing purposes.

Overall, the relationship between generative AI and ChatGPT is that ChatGPT is an implementation of generative AI that is specifically designed for conversational interactions. As a result, ChatGPT is a powerful tool for natural language processing that can be used in a wide range of applications, from customer service to education to healthcare.

As an AI language model, ChatGPT's future is constantly evolving and growing. With a temperature of 0.9, ChatGPT has the potential to generate more imaginative and unexpected responses, albeit at the cost of potentially introducing errors and inconsistencies.

In the future, ChatGPT will likely continue to improve its natural language processing capabilities, allowing it to understand and respond to increasingly complex and nuanced queries. It may also become more personalized, utilizing data from users' interactions to tailor responses to individual preferences and needs.

Additionally, ChatGPT may be used in a wide range of industries and applications, such as customer service, education, and healthcare, providing users with instant access to information and support.

However, as with any emerging technology, ChatGPT will face challenges, such as ethical concerns surrounding its use, potential biases in its responses, and the need to ensure user privacy and security.

Overall, the future of ChatGPT is exciting and full of potential. With continued development and improvement, ChatGPT has the potential to revolutionize the way we interact with technology and each other, making communication faster, more efficient, and more personalized.

Certainly! As an AI language model, ChatGPT has the potential to be used in a wide range of applications and industries. Here are some potential areas where ChatGPT could be used in the future:

1. **Customer service:** ChatGPT could be used by companies to provide instant and personalized support to their customers. Chatbots powered by ChatGPT could answer frequently asked questions, troubleshoot technical issues, and provide personalized recommendations to users.
2. **Education:** ChatGPT could be used in online learning environments to provide instant feedback and support to students. Chatbots powered

by ChatGPT could answer students' questions, provide personalized feedback on assignments, and help students navigate complex topics.

3. **Healthcare:** ChatGPT could be used in telemedicine applications to provide patients with instant access to medical advice and support. Chatbots powered by ChatGPT could answer patients' questions, provide guidance on medication regimens, and help patients track their symptoms.

4. **Journalism:** ChatGPT could be used in newsrooms to help journalists quickly gather and analyze information on breaking news stories. Chatbots powered by ChatGPT could scan social media and other sources for relevant information, summarize key points, and help journalists identify potential angles for their stories.

5. **Personalized marketing:** ChatGPT could be used by marketers to provide personalized recommendations and support to customers. Chatbots powered by ChatGPT could analyze users' browsing history, purchase history, and other data to provide personalized product recommendations and marketing messages.

Of course, as with any emerging technology, ChatGPT will face challenges and limitations. Some potential issues include:

1. **Ethical concerns:** There are ethical concerns surrounding the use of AI language models like ChatGPT, particularly with regard to issues like privacy, bias, and the potential for misuse.

2. **Accuracy and reliability:** ChatGPT is only as good as the data it is trained on, and it may not always provide accurate or reliable information. Ensuring that ChatGPT is trained on high-quality data and that its responses are validated and verified will be crucial to its success.

3. **User experience:** Ensuring that users have a positive and seamless experience interacting with ChatGPT will be crucial to its adoption and success. This may require improvements in natural language processing and user interface design.

Overall, the future of ChatGPT is full of potential and promise. With continued development and improvement, ChatGPT has the potential to transform the way we interact with technology and each other, making communication faster, more efficient, and more personalized than ever before.

8

Intellectual Abilities of Artificial Intelligence (AI)

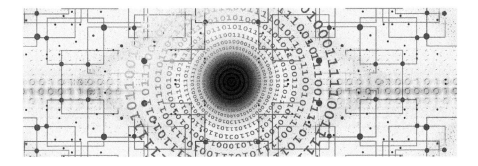

To understand AI's capabilities and abilities, we need to recognize the different components and subsets of AI. Terms like neural networks, machine learning (ML), and deep learning need to be defined and explained.

In general, artificial intelligence (AI) refers to the simulation of human intelligence in machines that are programmed to think like humans and mimic their actions. The term may also be applied to any machine that exhibits traits associated with a human mind such as learning and problem-solving [1].

Neural Networks

In information technology, a neural network is a system of programs and data structures that approximates the operation of the human brain. A neural network usually involves a large number of processors operating in parallel, each with its own small sphere of knowledge and access to data in its local memory.

Typically, a neural network is initially "trained" or fed large amounts of data and rules about data relationships (for example, "A grandfather is older than a person's father"). A program can then tell the network how to behave in response to an external stimulus (for example, to input from a computer user who is interacting with the network) or can initiate activity on its own (within the limits of its access to the external world).

Deep learning vs. Machine Learning (Figure 8.1)

To understand what deep learning is, it is first important to distinguish it from other disciplines within the field of AI.

One outgrowth of AI was machine learning, in which the computer extracts knowledge through supervised experience. This typically involved a human operator helping the machine learn by giving it hundreds or thousands of training examples, and manually correcting its mistakes.

While machine learning has become dominant within the field of AI, it does have its problems. For one thing, it is massively time consuming. For

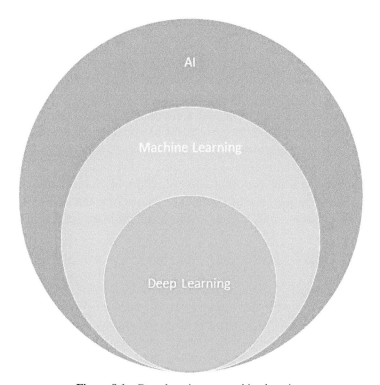

Figure 8.1   Deep learning vs. machine learning.

another, it is still not a true measure of machine intelligence since it relies on human ingenuity to come up with the abstractions that allow a computer to learn.

Unlike machine learning, deep learning is mostly *unsupervised*. It involves, for example, creating large-scale neural nets that allow the computer to learn and "think" by itself — without the need for direct human intervention.

Deep learning "really does not look like a computer program" where ordinary computer code is written in very strict logical steps, but what you will see in deep learning is something different; you do not have a lot of instructions that say: "If one thing is true do this other thing."

Instead of linear logic, deep learning is based on theories of how the human brain works. The program is made of tangled layers of interconnected nodes. It learns by rearranging connections between nodes after each new experience.

Deep learning has shown potential as the basis for software that could work out the emotions or events described in text (even if they are not explicitly referenced), recognize objects in photos, and make sophisticated predictions about people's likely future behavior. An example of deep learning in action is voice recognition like Google Now and Apple's Siri.

Deep learning is showing a great deal of promise — and it will make self-driving cars and robotic butlers a real possibility. The ability to analyze massive datasets and use deep learning in computer systems that can adapt to experience, rather than depending on a human programmer, will lead to breakthroughs. These range from drug discovery to the development of new materials to robots with a greater awareness of the world around them [2−8].

Deep Learning and Affective Computing

Affective computing is the study and development of systems and devices that can recognize, interpret, process, and simulate human affects. It is an interdisciplinary field spanning computer science (deep learning), psychology, and cognitive science. While the origins of the field may be traced as far back as to early philosophical inquiries into emotion ("affect" is, basically, a synonym for "emotion"), the more modern branch of computer science originated with *Rosalind Picard's* 1995 paper on affective computing. A motivation for the research is the ability to simulate *empathy*. The machine should interpret the emotional state of humans and adapt its behavior to them, giving an appropriate response for those emotions [9−12].

Affective computing technologies using deep learning sense the emotional state of a user (via sensors, microphone, cameras, and/or software logic) and respond by performing specific, predefined product/service features, such as changing a quiz or recommending a set of videos to fit the mood of the learner.

The more computers we have in our lives, the more we are going to want them to behave politely and be socially smart. We do not want it to bother us with unimportant information. That kind of common-sense reasoning requires an understanding of the person's *emotional state*.

One way to look at affective computing is *human−computer interaction* in which a device has the ability to detect and appropriately respond to its user's emotions and other stimuli. A computing device with this capacity could gather cues to user emotion from a variety of sources. Facial expressions, posture, gestures, speech, the force, or rhythm of key strokes and the temperature changes of the hand on a mouse can all signify changes in the user's emotional state, and these can all be detected and interpreted by a computer. A built-in camera captures images of the user and algorithms are used to process the data to yield meaningful information. Speech recognition and gesture recognition are among the other technologies being explored for affective computing applications.

Recognizing emotional information requires the extraction of meaningful patterns from the gathered data. This is done using deep learning techniques that process different modalities, such as speech recognition, natural language processing, or facial expression detection.

Emotion in Machines

A major area in affective computing is the design of computational devices proposed to exhibit either innate emotional capabilities or that are capable of convincingly simulating emotions. A more practical approach, based on current technological capabilities, is the simulation of emotions in conversational agents in order to enrich and facilitate interactivity between human and machine. While human emotions are often associated with surges in hormones and other neuropeptides, emotions in machines might be associated with abstract states associated with progress (or lack of progress) in autonomous learning systems; in this view, affective emotional states correspond to time-derivatives in the learning curve of an arbitrary learning system.

Two major categories describing emotions in machines: *emotional speech* and *facial affect detection*.

Emotional speech includes:

- Deep learning
- Databases
- Speech descriptors

Facial affect detection includes:

- Body gesture
- Physiological monitoring

The Future

Affective computing using deep learning tries to address one of the major drawbacks of online learning versus in-classroom learning – the teacher's capability to immediately adapt the pedagogical situation to the emotional state of the student in the classroom. In e-learning applications, affective computing using deep learning can be used to adjust the presentation style of a computerized tutor when a learner is bored, interested, frustrated, or pleased. Psychological health services, i.e., counseling, benefit from affective computing applications when determining a client's emotional state.

Robotic systems capable of processing affective information exhibit higher flexibility while one works in uncertain or complex environments. Companion devices, such as digital pets, use affective computing with deep learning abilities to enhance realism and provide a higher degree of autonomy.

Other potential applications are centered around social monitoring. For example, a car can monitor the emotion of all occupants and engage in additional safety measures, such as alerting other vehicles if it detects the driver to be angry. Affective computing with deep learning at the core has potential applications in human – computer interaction, such as affective mirrors allowing the user to see how he or she performs; emotion monitoring agents sending a warning before one sends an angry email; or even music players selecting tracks based on mood. Companies would then be able to use affective computing to infer whether their products will or will not be well received by the respective market. There are endless applications for affective computing with deep learning in all aspects of life.

9

Narrow AI vs. General AI vs. Super AI

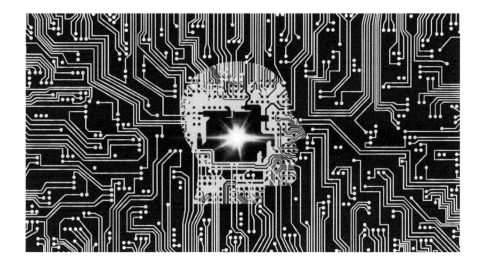

Artificial intelligence (AI) is a term used to describe machines that can perform tasks that normally require human intelligence, such as visual perception, speech recognition, decision-making, and language translation. AI is classified into three main types: narrow AI, general AI, and super AI. Each type of AI has its unique characteristics, capabilities, and limitations. In this article, we will explain the differences between these three types of AI.

Narrow AI

Narrow AI, also known as weak AI, refers to AI that is designed to perform a specific task or a limited range of tasks. It is the most common type of AI and is widely used in various applications such as facial recognition, speech recognition, image recognition, natural language processing, and recommendation systems.

Narrow AI works by using machine learning algorithms, which are trained on a large amount of data to identify patterns and make predictions. These algorithms are designed to perform specific tasks, such as identifying objects in images or translating languages. Narrow AI is not capable of generalizing beyond the tasks for which it is programmed, meaning that it cannot perform tasks that it has not been specifically trained to do.

One of the key advantages of narrow AI is its ability to perform tasks faster and more accurately than humans. For example, facial recognition systems can scan thousands of faces in seconds and accurately identify individuals. Similarly, speech recognition systems can transcribe spoken words with high accuracy, making it easier for people to interact with computers.

However, narrow AI has some limitations. It is not capable of reasoning or understanding the context of the tasks it performs. For example, a language translation system can translate words and phrases accurately, but it cannot understand the meaning behind the words or the cultural nuances that may affect the translation. Similarly, image recognition systems can identify objects in images, but they cannot understand the context of the images or the emotions conveyed by the people in the images.

General AI

General AI, also known as strong AI, refers to AI that is designed to perform any intellectual task that a human can do. It is a theoretical form of AI that is not yet possible to achieve. General AI would be able to reason, learn, and understand complex concepts, just like humans.

The goal of general AI is to create a machine that can think and learn in the same way that humans do. It would be capable of understanding language, solving problems, making decisions, and even exhibiting emotions. General AI would be able to perform any intellectual task that a human can do, including tasks that it has not been specifically trained to do.

One of the key advantages of general AI is that it would be able to perform any task that a human can do, including tasks that require creativity, empathy, and intuition. This would open up new possibilities for AI applications in fields such as healthcare, education, and the arts.

However, general AI also raises some concerns. The development of general AI could have significant ethical implications, as it could potentially surpass human intelligence and become a threat to humanity. It could also

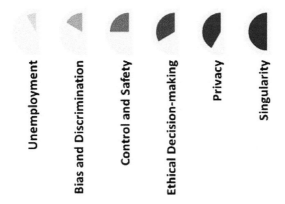

Figure 9.1   Challenges and ethical implications of general AI and super AI.

lead to widespread unemployment, as machines would be able to perform tasks that were previously done by humans. Here are a few examples of general AI:

1. AlphaGo: A computer program developed by Google's DeepMind that is capable of playing the board game Go at a professional level.

2. Siri: An AI-powered personal assistant developed by Apple that can answer questions, make recommendations, and perform tasks such as setting reminders and sending messages.

3. ChatGPT: A natural language processing tool driven by AI technology that allows you to have human-like conversations and much more with a chatbot. The language model can answer questions and assist you with tasks such as composing emails, essays, and code.

Super AI

Super AI refers to AI that is capable of surpassing human intelligence in all areas. It is a hypothetical form of AI that is not yet possible to achieve. Super AI would be capable of solving complex problems that are beyond human capabilities and would be able to learn and adapt at a rate that far exceeds human intelligence.

The development of super AI is the ultimate goal of AI research. It would have the ability to perform any task that a human can do, and more. It could potentially solve some of the world's most pressing problems, such as climate change, disease, and poverty.

Possible examples from movies: Skynet (Terminator), Viki (iRobot), and Jarvis (Ironman).

Challenges and Ethical Implications of General AI and Super AI

The development of general AI and super AI poses significant challenges and ethical implications for society. Some of these challenges and implications are discussed below (Figure 9.1).

1. **Control and safety:** General AI and super AI have the potential to become more intelligent than humans, and their actions could be difficult to predict or control. It is essential to ensure that these machines are safe and do not pose a threat to humans. There is a risk that these machines could malfunction or be hacked, leading to catastrophic consequences.
2. **Bias and discrimination:** AI systems are only as good as the data they are trained on. If the data is biased, the AI system will be biased as well. This could lead to discrimination against certain groups of people, such as women or minorities. There is a need to ensure that AI systems are trained on unbiased and diverse data.
3. **Unemployment:** General AI and super AI have the potential to replace humans in many jobs, leading to widespread unemployment. It is essential to ensure that new job opportunities are created to offset the job losses caused by these machines.
4. **Ethical decision-making:** AI systems are not capable of ethical decision-making. There is a need to ensure that these machines are programmed to make ethical decisions and that they are held accountable for their actions.
5. **Privacy:** AI systems require vast amounts of data to function effectively. This data may include personal information, such as health records and financial data. There is a need to ensure that this data is protected and that the privacy of individuals is respected.
6. **Singularity:** Some experts have raised concerns that general AI or super AI could become so intelligent that they surpass human intelligence, leading to a singularity event. This could result in machines taking over the world and creating a dystopian future.

Narrow AI, general AI, and super AI are three different types of AI with unique characteristics, capabilities, and limitations. While narrow AI is already in use in various applications, general AI and super AI are still theoretical and pose significant challenges and ethical implications. It is essential to ensure that AI systems are developed ethically and that they are designed to benefit the society as a whole.

10

Understanding the Psychological Impacts of using AI

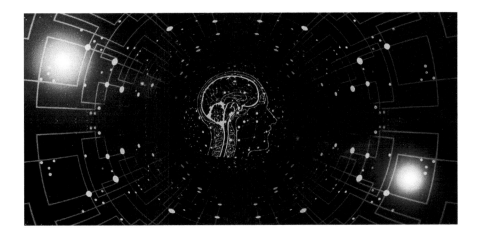

Artificial intelligence (AI) has rapidly become integrated into many aspects of our daily lives, from personal assistants on our smartphones to the algorithms that underpin social media feeds. While AI has the potential to revolutionize the way we live and work, it is not without its drawbacks. One area of concern is the potential *psychological impacts of using AI systems*. As we become increasingly reliant on these technologies, there is growing concern that they may be having negative effects on our mental health and well-being. In this article, we will explore the potential psychological impacts of using AI systems and discuss strategies for minimizing these risks.

The potential psychological impacts of using AI (Figure 10.1):

- **Anxiety:** Some people may feel anxious when using AI systems because they are not sure how the system works or what outcomes to expect. For example, if someone is using a speech recognition system to transcribe

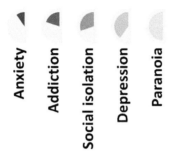

Figure 10.1   The potential psychological impacts of using AI.

their voice, they may feel anxious if they are unsure if the system is accurately capturing their words.

- **Addiction:** Overuse of technology, including AI systems, can lead to addictive behaviors. People may feel compelled to constantly check their devices or use AI-powered apps, which can interfere with other aspects of their lives, such as work or social relationships.
- **Social isolation:** People who spend too much time interacting with AI systems may become socially isolated, as they may spend less time engaging with other people in person. This can lead to a reduced sense of community or connection to others.
- **Depression:** Some people may experience depression or a sense of helplessness when interacting with AI systems that they perceive as being superior or more capable than they are. For example, if someone is using an AI-powered personal assistant, they may feel inadequate or helpless if the system is better at completing tasks than they are.
- **Paranoia:** Concerns around the safety and security of AI systems, as well as fears of AI taking over or replacing human decision-making, can lead to paranoid thinking in some individuals. This is particularly true in cases where AI systems are used to control physical systems, such as autonomous vehicles or weapons systems.

It is important to note that *not everyone* will experience these negative psychological impacts when using AI systems, and many people find AI to be helpful and beneficial. However, it is important to be aware of the potential risks associated with technology use and to take steps to mitigate these risks when possible.

There are several steps that you can take to minimize the potential negative psychological impacts of using AI systems:

- **Set boundaries:** Establish clear boundaries for your use of AI systems and try to limit your exposure to them. This can help prevent addiction and reduce feelings of anxiety or depression.
- **Stay informed:** Keep up-to-date with the latest developments in AI technology and try to understand how AI systems work. This can help reduce feelings of helplessness or paranoia and increase your confidence in using these systems.
- **Seek support:** If you are feeling anxious or stressed when using AI systems, talk to a trusted friend, family member, or mental health professional. They can provide support and help you work through your feelings.
- **Use AI systems responsibly:** When using AI systems, be mindful of their limitations and potential biases. Avoid relying solely on AI-generated information and always seek out multiple sources when making important decisions.
- **Take breaks:** Make sure to take regular breaks from using AI systems and spend time engaging in activities that promote relaxation and social connection. This can help reduce feelings of isolation and prevent addiction.
- **Advocate for ethical use of AI:** Support efforts to ensure that AI systems are developed and deployed in an ethical manner, with appropriate safeguards in place to protect privacy, autonomy, and other important values.

By following these steps, you can help ensure that your use of AI systems is positive and does not have negative psychological impacts.

Examples of incidents of psychological impacts of using AI:

- **AI-generated deepfake videos:** Deepfakes are videos that use AI to manipulate or replace an individual's image or voice in a video or audio recording. These videos can be used to spread false information or malicious content, which can have a severe psychological impact on the person depicted in the video.
- **Social media algorithms:** Social media platforms use AI algorithms to personalize the user experience by showing users content they are likely to engage with. However, this can create echo chambers where users only see content that aligns with their views, leading to confirmation bias and potentially increasing political polarization.
- **Job automation:** AI-powered automation can lead to job loss or significant changes in job roles and responsibilities. This can create anxiety

and stress for employees who fear losing their jobs or having to learn new skills.

- **Bias in AI algorithms:** AI algorithms can perpetuate bias and discrimination, particularly in areas like criminal justice or hiring. This can harm marginalized groups and lead to feelings of injustice and discrimination.
- **Dependence on AI:** As people become increasingly reliant on AI-powered tools and devices, they may experience anxiety or stress when they cannot access or use these tools.
- **Surveillance and privacy concerns:** AI-powered surveillance tools, such as facial recognition technology, can infringe on privacy rights and create a sense of unease or paranoia in individuals who feel like they are being constantly monitored.
- **Mental health chatbots:** AI-powered chatbots have been developed to provide mental health support and guidance to individuals. While these tools can be helpful for some people, they can also lead to feelings of isolation and disconnection if users feel like they are not receiving personalized or empathetic support.
- **Addiction to technology:** With the increasing prevalence of AI-powered devices, people may become addicted to technology, leading to symptoms such as anxiety, depression, and sleep disorders.
- **Virtual assistants:** Virtual assistants, such as Siri or Alexa, can create a sense of dependency and make it harder for individuals to engage in real-life social interactions.

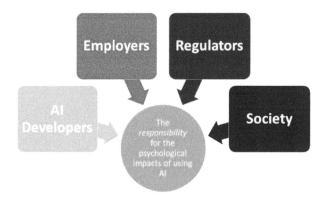

Figure 10.2   The responsibility for the psychological impacts of using AI.

- **Gaming and virtual reality:** AI-powered gaming and virtual reality experiences can create a sense of immersion and escapism, potentially leading to addiction and detachment from real-life experiences.

The *responsibility* for the psychological impacts of using AI falls on various individuals and organizations, including (Figure 10.2):

- **AI developers:** The developers of AI systems are responsible for ensuring that their systems are designed and programmed in a way that minimizes any negative psychological impacts on users. This includes considering factors such as transparency, privacy, and trustworthiness.
- **Employers:** Employers who use AI in the workplace have a responsibility to ensure that their employees are not negatively impacted by the use of AI. This includes providing training and support to help employees adjust to working with AI, as well as monitoring for any negative psychological impacts.
- **Regulators:** Government agencies and other regulatory bodies have a responsibility to ensure that the use of AI does not have negative psychological impacts on individuals. This includes setting standards and regulations for the design, development, and use of AI systems.
- **Society as a whole:** Finally, society as a whole has a responsibility to consider the psychological impacts of AI and to advocate for the development and use of AI systems that are designed with the well-being of individuals in mind. This includes engaging in public dialogue and debate about the appropriate use of AI, as well as advocating for policies that protect the rights and well-being of individuals impacted by AI.

Overall, the psychological impacts of AI are complex and multifaceted, and more research is needed to fully understand their effects.

11

Ethics in AI

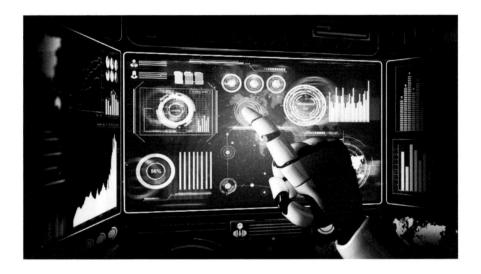

Artificial intelligence (AI) is transforming our world in countless ways, from healthcare to education, business to cybersecurity. While the potential benefits of AI are vast, there are also significant ethical considerations that must be taken into account. As intelligent machines become more prevalent in our society, it is crucial to consider the ethical implications of their use. In this essay, I will explore some of the key ethical considerations in AI, including *bias, privacy, accountability, and transparency* (Figure 11.1).

Bias in AI

One of the most significant ethical considerations in AI is *bias*. Bias can occur in AI systems when the data used to train them is biased or when the algorithms used to make decisions are biased. For example, facial recognition

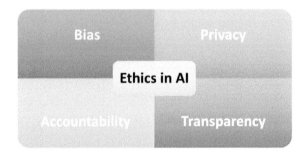

Figure 11.1 Ethics in AI.

systems have been shown to be less accurate in identifying people with darker skin tones. This is because the data used to train these systems was primarily made up of images of lighter-skinned individuals. As a result, the system is more likely to misidentify someone with darker skin.

Bias in AI can have serious consequences, particularly in areas like healthcare and criminal justice. For example, if an AI system is biased against certain groups of people, it could lead to inaccurate diagnoses or unequal treatment. To address this issue, it is essential to ensure that the data used to train AI systems is diverse and representative of the entire population. Additionally, AI systems should be regularly audited to detect and correct any biases that may arise.

Privacy in AI

Another ethical consideration in AI is *privacy*. As AI systems become more prevalent, they are collecting and processing vast amounts of data about individuals. This data can include everything from personal information like names and addresses to sensitive information like medical records and financial information. It is essential to ensure that this data is protected and used only for its intended purpose.

One of the biggest risks to privacy in AI is the potential for data breaches. If an AI system is hacked or otherwise compromised, it could lead to the exposure of sensitive information. To mitigate this risk, it is crucial to ensure that AI systems are designed with security in mind. Additionally, individuals should be given control over their data and should be able to choose whether or not it is collected and used by AI systems.

Accountability in AI

As AI systems become more autonomous, it is crucial to consider the issue of *accountability*. If an AI system makes a mistake or causes harm, who is responsible? The answer to this question is not always clear, particularly in cases where AI systems are making decisions that have significant consequences. For example, if an autonomous vehicle causes an accident, who is responsible? The manufacturer of the vehicle? The owner of the vehicle? The AI system itself?

To address this issue, it is essential to establish clear lines of accountability for AI systems. This could involve requiring manufacturers to take responsibility for the actions of their AI systems or establishing regulations that hold AI systems to a certain standard of safety and performance.

Transparency in AI

Finally, *transparency* is another critical ethical consideration in AI. As AI systems become more prevalent in our society, it is essential to ensure that they are transparent and understandable. This means that individuals should be able to understand how AI systems are making decisions and why they are making those decisions. Additionally, AI systems should be auditable, meaning that their decision-making processes can be reviewed and evaluated.

Transparency is particularly important in areas like healthcare and criminal justice, where decisions made by AI systems can have significant consequences. For example, if an AI system is used to make medical diagnoses, patients should be able to understand how the system arrived at its diagnosis and why that diagnosis was made. Similarly, if an AI system is used to make decisions about criminal sentencing, defendants should be able to understand how the system arrived at its decision and why that decision was made.

Ethical considerations in AI are crucial for ensuring that the technology is developed and used in a responsible and beneficial manner. As AI continues to advance and become more integrated into our daily lives, it is essential that we prioritize ethical considerations such as transparency, accountability, fairness, privacy, and safety. By doing so, we can harness the full potential of AI while mitigating any negative consequences. It is important for all stakeholders, including governments, industry leaders, researchers, and the general public, to engage in ongoing discussions and collaboration to establish ethical guidelines and best practices for the development and use of AI. Ultimately, a human-centric approach to AI ethics can help to ensure that AI is aligned with our values and benefits society as a whole.

Part II

AI Applications

12

AI and Robotics

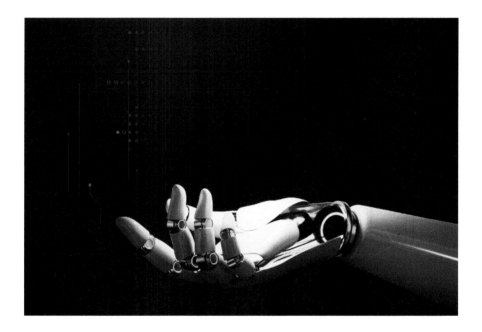

Artificial intelligence (AI) and robotics are two rapidly evolving fields that have the potential to revolutionize the way we live and work. In recent years, advances in these areas have been breathtaking, and there is no doubt that they will continue to shape the future of humanity. In this chapter, we will explore the future of AI and robots, looking at the current trends, challenges, and opportunities.

One of the most significant trends in AI and robotics is the growth of *machine learning*. This has led to significant advances in areas such as image recognition, natural language processing, and decision-making.

Another trend is the integration of AI and robotics into various industries, including: *healthcare and transportation* are two areas where these technologies are expected to have a significant impact. However, they are also being used in fields such as manufacturing, agriculture, and even space exploration.

In terms of challenges, one of the biggest concerns is *job displacement*. As machines become more capable of performing tasks traditionally done by humans, many jobs will become obsolete. This could lead to significant economic and social disruption if not handled carefully. However, there is also the potential for new types of jobs to emerge, and for industries to be created around the development and maintenance of these technologies.

Another challenge is the *ethical implications* of AI and robotics. As machines become more intelligent, they will need to make decisions that affect human lives. This raises important questions about the responsibility of those who design and operate these machines, as well as the potential for machines to be used for malicious purposes. It is essential that we have an open and transparent dialogue about these issues and that we work to ensure that these technologies are developed and used in a way that benefits humanity as a whole.

There are many opportunities for AI and robotics to improve our lives. From healthcare and transportation to education and entertainment, these technologies have the potential to make many aspects of our lives easier, safer, and more enjoyable. However, it is up to us to ensure that we use these technologies in a way that is responsible and equitable and that we continue to push the boundaries of what is possible.

The advancement of robotics and AI technology is rapidly changing the job market. Many jobs that were once considered safe from automation are now at risk of being replaced by smart robots. Following are some examples of jobs that could be replaced by smart robots:

1. **Manufacturing jobs:** Robots have already taken over many manufacturing jobs, such as those in automotive production lines, and are increasingly being used in other manufacturing sectors.
2. **Delivery and transportation jobs:** With the development of self-driving cars and drones, many delivery and transportation jobs may become obsolete.
3. **Customer service jobs:** Chatbots and virtual assistants are increasingly being used to handle customer service inquiries, which could eventually lead to the replacement of human customer service representatives.

4. **Data entry jobs:** The use of optical character recognition (OCR) technology is making it easier to digitize paper documents and automate data entry tasks.
5. **Accounting and bookkeeping jobs:** AI-powered software can now automate many of the tasks traditionally done by accountants and bookkeepers, such as data entry and reconciling accounts.
6. **Banking jobs:** AI-powered chatbots and virtual assistants are being used to handle routine banking inquiries, which could eventually lead to the replacement of human bank tellers.
7. **Medical jobs:** Robots are increasingly being used in surgical procedures, and AI-powered software is being used to diagnose diseases and develop treatment plans.
8. **Security jobs:** Drones and security robots are being used to patrol and monitor large areas, which could eventually lead to the replacement of human security guards.
9. **Cleaning jobs:** Robots are being developed to perform a range of cleaning tasks, from vacuuming floors to cleaning windows.
10. **Farming jobs:** With the development of autonomous tractors and drones, many farming tasks could be automated in the near future.

It is important to note that while these jobs may be at risk of automation, new jobs will also be created in fields such as robotics engineering and software development. The key is to ensure that workers are equipped with the skills needed to adapt to the changing job market.

Artificial intelligence (AI) and robots are changing the way we live and work. While they offer many benefits, there are also significant drawbacks to consider. We will explore the advantages and disadvantages of AI and robots.

Advantages of AI and Robots (Figure 12.1)

1. **Increased efficiency and productivity:** One of the most significant advantages of AI and robots is their ability to perform tasks faster and more efficiently than humans. This can lead to increased productivity and lower costs for businesses.
2. **Improved safety:** Robots can be used to perform dangerous or hazardous tasks, such as exploring deep-sea environments or cleaning up nuclear waste. This can help keep humans out of harm's way and reduce the risk of accidents.

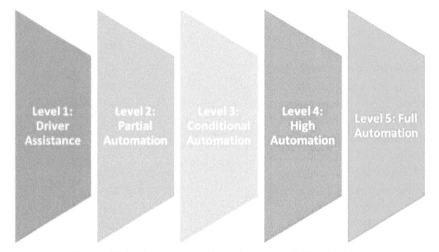

Figure 12.1   Advantages and disadvantages of AI and robots.

3. **Cost savings:** By using robots and AI, businesses can reduce labor costs and increase efficiency. This can lead to significant cost savings over time.
4. **Improved accuracy:** AI and robots can perform tasks with greater precision and accuracy than humans. This is particularly important in fields such as medicine and manufacturing, where small mistakes can have significant consequences.
5. **Greater access to information:** AI can be used to analyze vast amounts of data, making it easier for businesses and individuals to make informed decisions.

Disadvantages of AI and Robots (Figure 12.1)

1. **Job displacement:** One of the most significant drawbacks of AI and robots is the potential for job displacement. As machines become more capable of performing tasks traditionally done by humans, many jobs may become obsolete.
2. **Cost:** While AI and robots can lead to cost savings over time, the initial investment can be significant. This can make it difficult for small businesses and individuals to adopt these technologies.
3. **Dependence on technology:** As we become more reliant on AI and robots, we may become less capable of performing tasks without their

assistance. This could have long-term consequences for our ability to solve problems and make decisions independently.

4. **Ethical concerns:** AI and robots are increasingly being used in areas such as healthcare and law enforcement, raising important ethical questions about how these technologies are developed and used.

5. **Lack of empathy:** While AI and robots can perform tasks with great precision, they lack the empathy and emotional intelligence of humans. This can be a significant drawback in fields such as healthcare, where patients may need emotional support as well as medical care.

AI and robots offer many advantages, from increased efficiency and productivity to improved safety and accuracy. However, there are also significant drawbacks to consider, including job displacement, cost, and ethical concerns. As we continue to develop and adopt these technologies, it is essential that we carefully consider both the benefits and drawbacks and work to ensure that they are used in a way that benefits humanity as a whole. By doing so, we can create a future where AI and robots coexist with humans in a way that benefits everyone.

Overall, the future of AI and robotics is both exciting and uncertain. We are living in a time of rapid technological change, and it is up to us to shape the future in a way that benefits everyone. By working together, we can create a future where machines and humans coexist and thrive, and where the potential of AI and robotics is fully realized.

13

AI and Autonomous Cars

Autonomous cars have been a topic of discussion for several years. They have been the subject of numerous research projects and experiments in the field of artificial intelligence (AI). Autonomous cars are vehicles that are capable of driving themselves without human intervention. These vehicles use various sensors, cameras, and other technologies to detect their surroundings and make decisions based on the data collected. In this chapter, we will explore the different types of autonomous cars, the challenges facing the industry, the opportunities that they present, and their future prospects.

Types of Autonomous Cars

Autonomous cars are classified into different levels based on their degree of automation. *The levels range from 0 to 5*, with 0 being a vehicle that requires human intervention at all times, and 5 being a vehicle that is fully automated and requires no human intervention (Figure 13.1).

Level 1: Driver Assistance: These vehicles are equipped with some automated features, such as lane-keeping assistance and adaptive cruise control. However, they still require human intervention.

Level 2: Partial Automation: These vehicles can perform some tasks without human intervention, such as accelerating, braking, and steering. However, they still require human supervision.

Level 3: Conditional Automation: These vehicles are capable of performing most driving tasks without human intervention in certain conditions, such as on a highway. However, they require human intervention in certain situations.

Level 4: High Automation: These vehicles can perform all driving tasks without human intervention in most situations. However, they still require human intervention in extreme conditions.

Level 5: Full Automation: These vehicles can perform all driving tasks without human intervention in all conditions.

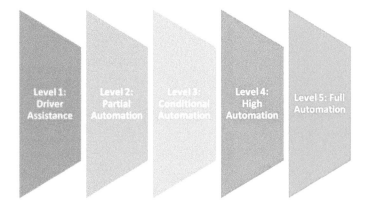

Figure 13.1   Types of autonomous cars.

Challenges Facing Autonomous Cars

Despite the significant progress made in the development of autonomous cars, there are still several challenges facing the industry. Some of these challenges include:

1. **Safety:** Safety is a major concern for the development of autonomous cars. The vehicles must be capable of detecting and responding to unexpected situations and must have fail-safe mechanisms in place to prevent accidents.
2. **Legal and regulatory challenges:** The development and deployment of autonomous cars are subject to legal and regulatory frameworks. These frameworks must be updated to account for the new technology, and the industry must navigate various legal and regulatory barriers to deployment.
3. **Cost:** The development and deployment of autonomous cars are expensive. The cost of sensors, cameras, and other technologies used in the vehicles can be prohibitive, making it difficult for manufacturers to produce affordable vehicles.

Opportunities Presented by Autonomous Cars

Autonomous cars present several opportunities, including:

1. **Improved safety:** Autonomous cars have the potential to significantly reduce accidents caused by human error, which account for a majority of accidents on the road.
2. **Increased efficiency:** Autonomous cars can optimize traffic flow, reducing congestion and travel time.
3. **Improved accessibility:** Autonomous cars can improve accessibility for people with disabilities and the elderly, providing them with greater mobility and independence.
4. **New business models:** Autonomous cars can enable new business models, such as ride-sharing and delivery services, which can reduce the need for car ownership and decrease the number of cars on the road.

Additional points to consider about AI and autonomous cars:

1. **Ethical considerations:** Autonomous cars must be programmed to make ethical decisions, such as in situations where they must choose between protecting the passengers and avoiding harm to others. This raises ethical considerations and challenges for the industry.

2. **Cybersecurity:** Autonomous cars rely heavily on software and connectivity to operate. This makes them vulnerable to cyberattacks, which can compromise their safety and security.
3. **Public perception:** The public's perception of autonomous cars is critical to their success. If the public perceives them as unsafe or unreliable, it could delay their adoption and slow down the industry's growth.
4. **Skillset and job loss:** The widespread adoption of autonomous cars could result in the loss of jobs for drivers and related professions, requiring the development of new skill sets and job opportunities.
5. **Environmental impact:** Autonomous cars have the potential to reduce emissions and improve air quality, especially if they are powered by clean energy sources.
6. **Data collection and privacy:** Autonomous cars collect vast amounts of data on their surroundings and passengers. This raises privacy concerns and the need for robust data collection and management policies.

Overall, the development and deployment of autonomous cars require collaboration between different stakeholders, including manufacturers, policymakers, and the public, to ensure their safety, efficiency, and accessibility.

The Future of Autonomous Cars

The future of autonomous cars is bright, with the industry projected to grow significantly in the coming years. According to recent reports, the global autonomous car market is expected to reach $556.67 billion by 2026, growing at a CAGR of 39.47% from 2019 to 2026. The growth is expected to be driven by advancements in AI and the increasing demand for safer and more efficient transportation.

Autonomous cars have the potential to revolutionize the transportation industry, providing safer, more efficient, and accessible transportation options. Despite the challenges facing the industry, the future of autonomous cars is bright, with significant growth expected in the coming years.

14

Quantum Computing and AI: A Transformational Match

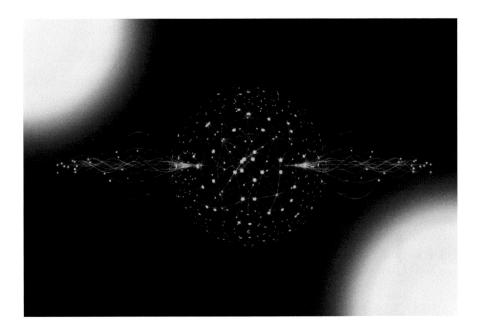

Quantum computers are designed to perform tasks much more accurately and efficiently than conventional computers, providing developers with a new tool for specific applications. It is clear in the short term that quantum computers will not replace their traditional counterparts; instead, they will require classical computers to support their specialized abilities, such as systems optimization [13].

Quantum computing and artificial intelligence are both transformational technologies and artificial intelligence needs quantum computing to achieve significant progress. Although artificial intelligence produces functional applications with classical computers, it is limited by the computational capabilities of classical computers. Quantum computing can provide a computation boost to artificial intelligence, enabling it to tackle more complex problems in many fields in business and science [16].

What is Quantum Computing?

Quantum computing is the area of study focused on developing computer technology based on the principles of quantum theory. The quantum computer, following the laws of quantum physics, would gain enormous processing power through the ability to be in multiple states, and to perform tasks using all possible permutations simultaneously.

A Comparison of Classical and Quantum Computing

Classical computing relies, at its ultimate level, on principles expressed by Boolean algebra. Data must be processed in an exclusive binary state at any point in time or bits. While the time that each transistor or capacitor need to be either in 0 or 1 before switching states is now measurable in billionths of a second, there is still a limit as to how quickly these devices can be made to switch state. As we progress to smaller and faster circuits, we begin to reach the physical limits of materials and the threshold for classical laws of physics to apply. Beyond this, the quantum world takes over. In a quantum computer, a number of elemental particles such as electrons or photons can be used with either their *charge* or *polarization* acting as a representation of 0 and/or 1. Each of these particles is known as a quantum bit, or *qubit*, and the nature and behavior of these particles form the basis of quantum computing.

Quantum Superposition and Entanglement

The two most relevant aspects of quantum physics are the principles of *superposition* and *entanglement*.

Superposition: Think of a qubit as an electron in a magnetic field. The electron's spin may be either in alignment with the field, which is known as a spin-up state, or opposite to the field, which is known as a spin-down state. According to quantum law, the particle enters a superposition of states, in which it behaves as if it were in both states simultaneously. Each qubit utilized could take a superposition of both 0 and 1.

Entanglement: Particles that have interacted at some point retain a type of connection and can be entangled with each other in pairs, in a process known as *correlation*. Knowing the spin state of one entangled particle — up or down — allows one to know that the spin of its mate is in the opposite direction. Quantum entanglement allows qubits that are separated by incredible distances to interact with each other instantaneously (not limited to the speed of light). No matter how great the distance between the correlated particles, they will remain entangled as long as they are isolated. Taken together, quantum superposition and entanglement create an enormously enhanced computing power. Here, a 2-bit register in an ordinary computer can store only one of four binary configurations (00, 01, 10, or 11) at any given time, and a 2-qubit register in a quantum computer can store all four numbers *simultaneously*, because each qubit represents two values. If more qubits are added, the increased capacity is expanded exponentially [13].

Difficulties with Quantum Computers

1. **Interference:** During the computation phase of a quantum calculation, the slightest disturbance in a quantum system (say a stray photon or wave of EM radiation) causes the quantum computation to collapse, a process known as *de-coherence*. A quantum computer must be totally isolated from all external interference during the computation phase.
2. **Error correction:** Given the nature of quantum computing, error correction is ultra-critical — even a single error in a calculation can cause the validity of the entire computation to collapse.
3. **Output observance:** Closely related to the above two, retrieving output data after a quantum calculation is complete risks corrupting the data.

Applications of Quantum Computing and AI

Keeping in mind that the term "quantum AI" means the use of quantum computing for computation of machine learning algorithms, which takes advantage of computational superiority of quantum computing, to achieve results that are not possible to achieve with classical computers, the following are some of the applications of this super mix of quantum computing and AI [16] (Figure 14.1).

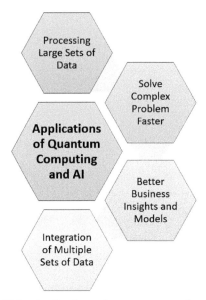

Figure 14.1   Applications of quantum computing and AI.

Processing Large Sets of Data

We produce 2.5 exabytes of data every day. That is equivalent to 250,000 Libraries of Congress or the content of 5 million laptops. Every minute of every day, 3.2 billion global internet users continue to feed the data banks with 9722 pins on Pinterest, 347,222 tweets, 4.2 million Facebook likes plus *all* the other data we create by taking pictures and videos, saving documents, opening accounts, and more [15].

Quantum computers are designed to manage the huge amount of data, along with uncovering patterns and spotting anomalies extremely quickly. With each newly launched iteration of quantum computer design and the new improvements made on the quantum error-correction code, developers are now able to better manage the potential of quantum bits. They also optimize the same for solving all kinds of business problems to make better decisions [15].

Solve Complex Problems Faster

Quantum computers can complete calculations within seconds, which would take today's computers many years to calculate. With quantum computing, developers can do multiple calculations with multiple inputs simultaneously. Quantum computers are critical to process the monumental amount of data that businesses generate on a daily basis, and the fast calculation can be used to solve very complex problems, which can be expressed as Quantum Supremacy where the calculations that normally take more than 10,000 years to perform, quantum computer can do it in *200 seconds*. The key is to translate real-world problems that companies are facing into quantum language [14,18]

Better Business Models

As the volume of data generated continues to surge in industries such as pharmaceuticals, finance, and life sciences, companies are increasingly distancing themselves from conventional computing methods. In pursuit of a more robust data framework, these organizations now necessitate intricate models that possess the processing power to effectively simulate even the most complex scenarios. This is precisely where quantum computers come into prominence. Harnessing the capabilities of quantum technology to construct more sophisticated models holds the promise of advancing disease treatments in the healthcare sector, particularly for ailments like COVID-19. Furthermore, optimizing the research cycle encompassing testing, tracing, and treatment of the virus has the potential to avert financial crises within the banking sector and enhance the efficiency of logistics chains in the manufacturing industry.

Integration of Multiple Sets of Data

To manage and integrate multiple numbers of sets of data from multiple sources, quantum computers are best to help, which makes the process quicker and also makes the analysis easier. The ability to handle so many stakes make quantum computing an adequate choice for solving business problems in a variety of fields [14].

The Future

The quantum computing market will reach $2.2 billion, and the number of installed quantum computers will reach around 180 in 2026, with about 45 machines produced in that year. These include both machines installed at the quantum computer companies themselves that are accessed by quantum services as well as customer premises machines [15].

Cloud access revenues will likely dominate as a revenue source for quantum computing companies in the format of Quantum Computing as a Service (QCaaS) offering, which will be accounting for 75% of all quantum computing revenues in 2026. Although in the long run quantum computers may be more widely purchased, today, potential end users are more inclined to do quantum computing over the cloud rather than making technologically risky and expensive investments in quantum computing equipment [17].

In parallel with the development of quantum software applications, the tools available to quantum developers and the pool of quantum engineers and experts will experience significant growth over the next five years. This expansion of infrastructure will enable a broader range of organizations to harness the potential of two transformative technologies: quantum computing and artificial intelligence (AI). Moreover, this trend is likely to encourage many universities to incorporate quantum computing as an integral component of their academic curriculum.

15

Blockchain and AI: A Perfect Match?

Blockchain and artificial intelligence are two of the hottest technology trends right now. Even though the two technologies have highly different developing parties and applications, researchers have been discussing and exploring their combination [24].

PwC predicts that by 2030, **AI will add up to $15.7 trillion to the world economy**, and as a result, global GDP will rise by 14%. According to Gartner's prediction, business value added by blockchain technology will increase to $3.1 trillion by the same year.

By definition, a blockchain is a distributed, decentralized, and immutable ledger used to store encrypted data. On the other hand, AI is the engine or the "brain" that will enable analytics and decision-making from the data collected [19].

It goes without saying that each technology has its own individual degree of complexity, but both AI and blockchain are in situations where they can benefit from each other and help one another [21].

With both these technologies being able to effect and enact upon data in different ways, their coming together makes sense, and it can take the exploitation of data to new levels. At the same time, the integration of machine learning and AI into blockchain, and vice versa, can enhance blockchain's underlying architecture and boost AI's potential [20].

Additionally, blockchain can also make AI more coherent and understandable, and we can trace and determine why decisions are made in machine learning. Blockchain and its ledger can record all data and variables that go through a decision made under machine learning.

Moreover, AI can boost blockchain efficiency far better than humans, or even standard computing can. A look at the way in which blockchains are currently run on standard computers proves this with a lot of processing power needed to perform even basic tasks [21].

Applications of AI and Blockchain (Figure 15.1)

Smart Computing Power

If you were to operate a blockchain, with all its encrypted data, on a computer, you would need large amounts of processing power. The hashing algorithms used to mine Bitcoin blocks, for example, take a "brute force" approach – which consists in systematically enumerating all possible candidates for the solution and checking whether each candidate satisfies the problem's statement before verifying a transaction [21].

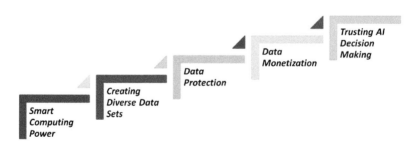

Applications of AI and Blockchain

Figure 15.1 Applications of AI and blockchain.

AI affords us the opportunity to move away from this and tackle tasks in a more intelligent and efficient way. Imagine a machine learning-based algorithm, which could practically polish its skills in "real-time" if it were fed the appropriate training data [21].

Creating Diverse Datasets

Unlike artificial-intelligence-based projects, blockchain technology creates decentralized, transparent networks that can be accessed by anyone, around the world in public blockchain networks situation. While blockchain technology is the ledger that powers cryptocurrencies, blockchain networks are now being applied to a number of industries to create decentralization. For example, SingularityNET is specifically focused on using blockchain technology to encourage a broader distribution of data and algorithms, helping ensure the future development of artificial intelligence and the creation of "decentralized AI" [22].

SingularityNET combines blockchain and AI to create smarter, decentralized AI blockchain networks that can host diverse datasets. By creating an API of APIs on the blockchain, it would allow for the intercommunication of AI agents. As a result, diverse algorithms can be built on diverse datasets [22].

Data Protection

The progress of AI is completely dependent on the input of data − our data. Through data, AI receives information about the world and things happening on it. Basically, data feeds AI, and through it, AI will be able to continuously improve itself.

On the other side, blockchain is essentially a technology that allows for the encrypted storage of data on a distributed ledger. It allows for the creation of fully secured databases that can be looked into by parties who have been approved to do so. When combining blockchains with AI, we have a backup system for the sensitive and highly valuable personal data of individuals.

Medical or financial data are too sensitive to hand over to a single company and its algorithms. Storing this data on a blockchain, which can be accessed by an AI, but only with permission and once it has gone through the proper procedures, could give us the enormous advantages of personalized recommendations while safely storing our sensitive data [22].

Data Monetization

Another disruptive innovation that could be possible by combining the two technologies is the monetization of data. Monetizing collected data is a huge revenue source for large companies, such as Facebook and Google [22].

Having others decide how data is being sold in order to create profits for businesses demonstrates that data is being weaponized against us. Blockchain allows us to cryptographically protect our data and have it used in the ways we see fit. This also lets us monetize data personally if we choose to, without having our personal information compromised. This is important to understand in order to combat biased algorithms and create diverse datasets in the future [22].

The same goes for AI programs that need our data. In order for AI algorithms to learn and develop, AI networks will be required to buy data directly from its creators, through data marketplaces. This will make the entire process a far more fair process than it currently is, without tech giants exploiting its users [22].

Such a data marketplace will also open up AI for smaller companies. Developing and feeding AI is incredibly costly for companies that do not generate their own data. Through decentralized data marketplaces, they will be able to access otherwise too expensive and privately kept data.

Trusting AI Decision-Making

As AI algorithms become smarter through learning, it will become increasingly difficult for data scientists to understand how these programs came to specific conclusions and decisions. This is because AI algorithms will be able to process incredibly large amounts of data and variables. However, we must continue to audit conclusions made by AI because we want to make sure they are still reflecting reality.

Through the use of blockchain technology, there are immutable records of all the data, variables, and processes used by AIs for their decision-making processes. This makes it far easier to audit the entire process.

With the appropriate blockchain programming, all steps from data entry to conclusions can be observed, and the observing party will be sure that this data has not been tampered with. It creates trust in the conclusions drawn by AI programs. This is a necessary step, as individuals and companies will not start using AI applications if they do not understand how they function, and on what information they base their decisions [23].

Conclusion

The combination of blockchain technology and artificial intelligence is still a largely undiscovered area. Even though the convergence of the two technologies has received its fair share of scholarly attention, projects devoted to this groundbreaking combination are still scarce.

Putting the two technologies together has the potential to use data in ways never before thought possible. Data is the key ingredient for the development and enhancement of AI algorithms, and blockchain secures this data, allows us to audit all intermediary steps that AI takes to draw conclusions from the data, and allows individuals to monetize their produced data.

AI can be incredibly revolutionary, but it must be designed with utmost precautions – blockchain can greatly assist in this. How the interplay between the two technologies will progress is anyone's guess. However, its potential for true disruption is clearly there and rapidly developing [24].

16

First Line of Defense for Cybersecurity: AI

The year 2017 was not a great year for cybersecurity; we saw a large number of high-profile cyberattacks, including Uber, Deloitte, Equifax, and the now infamous WannaCry ransomware attack, and 2018 started with a bang too with the hacking of Winter Olympics. The frightening truth about increasing cyberattacks is that most businesses and the cybersecurity industry itself are not prepared. Despite the constant flow of security updates and patches, the number of attacks continues to rise.

Beyond the lack of preparedness on the business level, the cybersecurity workforce itself is also having an incredibly hard time keeping up with demand. By the end of 2021, it is estimated that there will be an astounding 3.5 million unfilled cybersecurity positions worldwide. The current staff is overworked with an average of 52 hours a week, which is not an ideal situation to keep up with nonstop threats.

Given the state of cybersecurity today, the implementation of AI systems into the mix can serve as a real turning point. New AI algorithms use machine learning (ML) to adapt over time and make it easier to respond to cybersecurity risks. However, new generations of malware and cyberattacks can be difficult to detect with conventional cybersecurity protocols. They evolve over time; so more dynamic approaches are necessary.

Another great benefit of AI systems in cybersecurity is that they will free up an enormous amount of time for tech employees. Another way AI systems can help is by categorizing attacks based on the threat level. While there is still a fair amount of work to be done here, but when machine learning principles are incorporated into your systems, they can actually adapt over time, giving you a dynamic edge over cyber criminals.

Unfortunately, there will always be limits of AI, and human−machine teams will be the key to solving increasingly complex cybersecurity challenges. But as our models become effective at detecting threats, bad actors will look for ways to confuse the models. It is a field called adversarial machine learning, or adversarial AI. Bad actors will study how the underlying models work and work to either confuse the models − what experts call poisoning the models, or machine learning poisoning (MLP) – or focus on a wide range of evasion techniques, essentially looking for ways they can circumvent the models.

Four Fundamental Security Practices

With all the hype surrounding AI, we tend to overlook a very important fact. The best defense against a potential AI cyberattack is rooted in maintaining a fundamental security posture that incorporates continuous monitoring, user education, diligent patch management, and basic configuration controls to address vulnerabilities, which are explained in the following (Figure 16.1).

Identifying the Patterns

AI is all about patterns. Hackers, for example, look for patterns in server and firewall configurations, use of outdated operating systems, user actions and response tactics, and more. These patterns give them information about network vulnerabilities that they can exploit.

Network administrators also look for patterns. In addition to scanning for patterns in the way hackers attempt intrusions, they are trying to identify

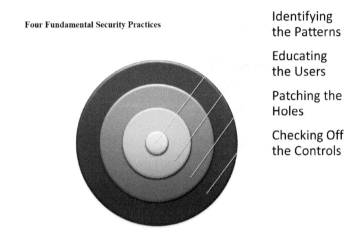

Four Fundamental Security Practices

Identifying
the Patterns

Educating
the Users

Patching the
Holes

Checking Off
the Controls

Figure 16.1   Four fundamental security practices.

potential anomalies like spikes in network traffic, irregular types of network traffic, unauthorized user logins and other red flags.

By collecting data and monitoring the state of their network under normal operating conditions, administrators can set up their systems to automatically detect when something unusual takes place – a suspicious network login, for example, or access through a known bad IP. This fundamental security approach has worked extraordinarily well in preventing more traditional types of attacks, such as malware or phishing. It can also be used very effectively in deterring AI-enabled threats.

Educating the Users

An organization could have the best monitoring systems in the world, but the work they do can all be undermined by a single employee clicking on the wrong email. Social engineering continues to be a large security challenge for businesses because workers easily can be tricked into clicking on suspicious attachments, emails, and links. Employees are considered by many as the *weakest links* in the security chain, as evidenced by a recent survey that found that careless and untrained insiders represented the top source of security threats.

Educating users on what not to do is just as important as putting security safeguards in place. Experts agree that routine user testing reinforces training.

Agencies must also develop plans that require all employees to understand their individual roles in the battle for better security. And do not forget a response and recovery plan; so everyone knows what to do and expect when a breach occurs. Test these plans for effectiveness. Do not wait for an exploit to find a hole in the process.

Patching the Holes

Hackers know when a patch is released, and in addition to trying to find a way around that patch, they will not hesitate to test if an agency has implemented the fix. Not applying patches opens the door to potential attacks − and if the hacker is using AI, those attacks can come much faster and be even more insidious.

Checking Off the Controls

The Center for Internet Security (CIS) has issued a set of controls designed to provide agencies with a checklist for better security implementations. While there are 20 actions in total, implementing at least the top five − device inventories, software tracking, security configurations, vulnerability assessments, and control of administrative privileges − can eliminate roughly 85% of an organization's vulnerabilities. All of these practices − monitoring, user education, patch management, and adherence to CIS controls − can help agencies fortify themselves against even the most sophisticated AI attacks.

Challenges Facing AI in Cybersecurity (Figure 16.2)

AI-powered Attacks

AI/machine learning (ML) software has the ability to "learn" from the consequences of past events in order to help predict and identify cybersecurity threats. According to a report by Webroot, AI is used by approximately 87% of US cybersecurity professionals. However, AI may prove to be a double-edged sword as 91% of security professionals are concerned that hackers will use AI to launch even more sophisticated cyberattacks.

For example, AI can be used to automate the collection of certain information − perhaps relating to a specific organization − which may be sourced from support forums, code repositories, social media platforms, and more. Additionally, AI may be able to assist hackers when it comes to cracking

Figure 16.2   Challenges facing AI in cybersecurity.

passwords by narrowing down the number of probable passwords based on geography, demographics, and other such factors.

More Sandbox-evading Malware

In recent years, sandboxing technology has become an increasingly popular method for detecting and preventing malware infections. However, cyber criminals are finding more ways to evade this technology. For example, new strains of malware are able to recognize when they are inside a sandbox, and wait until they are outside the sandbox before executing the malicious code.

Ransomware and IoT

We should be very careful not to underestimate the potential damage IoT ransomware could cause. For example, hackers may choose to target critical systems such as power grids. Should the victim fail to the pay the ransom within a short period of time, the attackers may choose to shut down the grid. Alternatively, they may choose to target factory lines, smart cars, and home appliances such as smart fridges, smart ovens, and more.

This fear was realized with a massive distributed denial of service attack that crippled the servers of services like Twitter, NetFlix, NYTimes, and PayPal across the U.S. on October 21, 2016. It is the result of an immense assault that involved millions of Internet addresses and malicious software, according to Dyn, the prime victim of that attack. "One source of the traffic for the attacks was devices infected by the Mirai botnet." The attack comes amid heightened cybersecurity fears and a rising number of internet security breaches. Preliminary indications suggest that countless Internet of Things (IoT) devices that power everyday technology like closed-circuit cameras and smart-home devices were hijacked by the malware and used against the servers.

A Rise of State-sponsored Attacks

The rise of nation state cyberattacks is perhaps one of the most concerning areas of cybersecurity. Such attacks are usually politically motivated and go beyond financial gain. Instead, they are typically designed to acquire intelligence that can be used to obstruct the objectives of a given political entity. They may also be used to target electronic voting systems in order to manipulate public opinion in some way.

As you would expect, state-sponsored attacks are targeted, sophisticated, and well-funded and have the potential to be incredibly disruptive. Of course, given the level of expertise and finance that is behind these attacks, they may prove very difficult to protect against. Governments must ensure that their internal networks are isolated from the internet and ensure that extensive security checks are carried out on all staff members. Likewise, staff will need to be sufficiently trained to spot potential attacks.

Shortage of Skilled Staff

By practically every measure, cybersecurity threats are growing more numerous and sophisticated each passing day, a state of affairs that does not bode well for an IT industry struggling with a security skills shortage. With less security talent to go around, there is a growing concern that businesses will lack the expertise to thwart network attacks and prevent data breaches in the years ahead.

IT Infrastructure

A modern enterprise has just too many IT systems, spread across geographies. Manual tracking of the health of these systems, even when they operate in a highly integrated manner, poses massive challenges. For most businesses, the only practical method of embracing advanced (and expensive) cybersecurity technologies is to prioritize their IT systems and cover those that they deem critical for business continuity. Currently, cybersecurity is reactive. That is to say that in most cases, it helps alert IT staff about data breaches, identity theft, suspicious applications, and suspicious activities. So, cybersecurity is currently more of an enabler of disaster management and mitigation. This leaves a crucial question unanswered – what about not letting cybercrime happen at all?

The Future of Cybersecurity and AI

In the security world, AI has a very clear-cut potential for good. The industry is notoriously unbalanced, with the bad actors getting to pick from thousands of vulnerabilities to launch their attacks, along with deploying an ever-increasing arsenal of tools to evade detection once they have breached a system. While they only have to be successful once, the security experts tasked with defending a system have to stop every attack, every time.

With the advanced resources, intelligence and motivation to complete an attack found in high level attacks, and the sheer number of attacks happening every day, victory eventually becomes impossible for the defenders.

The analytical speed and power of our dream security AI would be able to tip these scales at last, leveling the playing field for the security practitioners who currently have to constantly defend at scale against attackers who can pick a weak spot at their leisure. Instead, even the well-planned and concealed attacks could be quickly found and defeated.

Of course, such a perfect security AI is some way off. Not only would this AI need to be a bona fide simulated mind that can pass the Turing test, it would also need to be a fully trained cyber security professional, capable of replicating the decisions made by the most experienced security engineer, but on a vast scale.

Before we reach the brilliant AI seen in Sci-Fi, we need to go through some fairly testing stages – although these still have huge value in themselves. Some truly astounding breakthroughs are happening all the time. When it matures as a technology, it will be one of the most astounding developments in history, changing the human condition in ways similar to and bigger than electricity, flight, and the internet, because we are entering the AI era.

17

IoT, AI, and Blockchain: Catalysts for Digital Transformation

The digital revolution has brought with it a new way of thinking about manufacturing and operations. Emerging challenges associated with logistics and energy costs are influencing global production and associated distribution decisions. Significant advances in technology, including big data analytics, AI, Internet of Things, robotics, and additive manufacturing, are shifting the capabilities and value proposition of global manufacturing. In response, manufacturing and operations require a digital renovation: the value chain must be redesigned and retooled and the workforce retrained. Total delivered cost must be analyzed to determine the best places to locate sources of supply, manufacturing, and assembly operations around the world. In other words, we need a digital transformation.

Digital Transformation

Digital transformation (DX) is the profound transformation of business and organizational activities, processes, competencies, and models to fully leverage the changes and opportunities of a mix of digital technologies and their accelerating impact across society in a strategic and prioritized way, with present and future shifts in mind.

A digital transformation strategy aims to create the capabilities of fully leveraging the possibilities and opportunities of new technologies and their impact faster, better, and in a more innovative way in the future.

A digital transformation journey needs a staged approach with a clear road-map, involving a variety of stakeholders, beyond silos and internal/external limitations. This road-map takes into account that end goals will continue to move as digital transformation *de facto* is an ongoing journey, as is change and digital innovation.

Internet of Things (IoT)

IoT is defined as a system of interrelated *physical objects, sensors, actuators, virtual objects, people, services, platforms, and networks* that have separate identifiers and an ability to transfer data independently. Practical examples of IoT application today include precision agriculture, remote patient monitoring, and driverless cars. Simply put, IoT is the network of "things" that collects and exchanges information from the environment.

IoT is sometimes referred to as the driver of the *fourth industrial revolution* (Industry 4.0) by industry insiders and has triggered technological changes that span a wide range of fields. Gartner forecasted that there would be 20.8 billion connected things in use worldwide by 2020. IoT developments bring exciting opportunities to make our personal lives easier as well as improving efficiency, productivity, and safety for many businesses.

IoT and digital transformation are closely related for the following reasons:

1. More than 50% of companies think IoT is strategic, and one in four believes it is transformational.

2. Both increase company longevity. The average company's lifespan has decreased from 67 years in the 1920s to 15 years today.
3. One in three industry leaders will be digitally disrupted by 2020.
4. Both enable businesses to connect with customers and partners in open digital ecosystems, to share digital insights, collaborate on solutions, and share in the value created.
5. Competitors are doing it. According to IDC, 70% of global discrete manufacturers will offer connected products by 2023.
6. It is where the money is. Digital product and service sales are growing and will represent more than $1 of every $3 spent by 2021.
7. Enterprises are overwhelmed by data and digital assets. They already struggle to manage the data and digital assets they have, and IoT will expand them exponentially. They need help finding the insights in the vast stream of data and manage digital assets.
8. Both drive consumption. Digital services easily prove their own worth. Bundle products with digital services and content to make it easy for customers to consume them.
9. Both make companies understand customers better. Use integrated channels, big data, predictive analytics, and machine learning to uncover, predict, and meet customer needs, increasing loyalty and revenues, IoT, and AI are at the heart of this.
10. Utilizing both of these approaches ensures the future resilience of the business. By making well-informed strategic decisions for the company, its product and service portfolio, and future investments through the application of IoT data analytics, visualization, and AI, you can position the business for long-term success.

Digital Transformation, Blockchain, and AI

Digital transformation is a complicated challenge, but the integration of blockchain and AI makes it much easier. Considering the number of partners (internal, external, or both) involved in any given business process, a system in which a multitude of electronic parties can securely communicate, collaborate, and transact without human intervention is highly agile and efficient.

Enterprises that embrace this transformation will be able to provide a better user experience, a more consistent workflow, more streamlined operations and value-added services, as well as gain competitive advantage and differentiation.

Blockchain can holistically manage steps and relationships where participants will share the same data source, such as financial relationships and transactions connected to each step; security and accountability is factored in, as well as compliance with government regulations along with internal rules and processes. The result is consistency, reductions in costs and time delays, improved quality, and reduced risks.

AI can help companies learn in ways that accelerate innovation and can assist companies getting closer to customers and improve employee's productivity and engagement. Digital transformation efforts can be improved with that information.

The building blocks of digital transformation are: *mindset*, *people*, *process*, and *tools*. IoT covers all the blocks since IoT does not just connect devices, it connects people too. Blockchain will ensure end-to-end security and by using AI, you will move IoT beyond connections to intelligence. One important step is to team up with the best partners and invest in education, training, and certifying your teams. This magical mix of IoT, AI, and blockchain will help make transformation digital and easy.

18

AI is the Catalyst of IoT

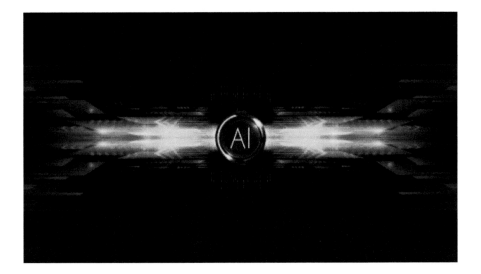

Businesses across the world are rapidly leveraging the Internet of Things (IoT) to create new products and services that are opening up new business opportunities and creating new business models. The resulting transformation is ushering in a new era of how companies run their operations and engage with customers. However, tapping into the IoT is only part of the story [30].

For companies to realize the full potential of IoT enablement, they need to combine IoT with rapidly advancing artificial intelligence (AI) technologies, which enable "smart machines" to simulate intelligent behavior and make well-informed decisions with little or no human intervention [30].

Artificial intelligence (AI) and the Internet of Things (IoT) are terms that project futuristic, Sci-Fi, imagery; both have been identified as drivers of business disruption in 2017. But, what do these terms really mean and what is their relation? Let us start by defining both terms first.

IoT is defined as a system of interrelated *physical objects, sensors, actuators, virtual objects, people, services, platforms, and networks* [3] that have separate identifiers and an ability to transfer data independently. Practical examples of IoT application today include precision agriculture, remote patient monitoring, and driverless cars. Simply put, IoT is the network of "things" that collects and exchanges information from the environment [31].

IoT is sometimes referred to as the driver of the *fourth industrial revolution* (Industry 4.0) by industry insiders and has triggered technological changes that span a wide range of fields. Gartner forecasted that there would be 20.8 billion connected things in use worldwide by 2020, but more recent predictions put the 2020 figure at over 50 billion devices [28]. Various other reports have predicted huge growth in a variety of industries, such as estimating healthcare IoT to be worth $117 billion by 2020 and forecasting 250 million connected vehicles on the road by the same year. IoT developments bring exciting opportunities to make our personal lives easier as well as improving efficiency, productivity, and safety for many businesses [26].

AI, on the other hand, is the engine or the "brain" that will enable analytics and decision-making from the data collected by IoT. In other words, IoT collects the data and AI processes this data in order to make sense of it. You can see these systems working together at a personal level in devices like fitness trackers and Google Home, Amazon's Alexa, and Apple's Siri [25].

With more connected devices comes more data that has the potential to provide amazing insights for businesses but presents a new challenge for how to analyze it all. Collecting this data benefits no one unless there is a way to understand it all. This is where AI comes in. Making sense of huge amounts of data is a perfect application for pure AI.

By applying the analytic capabilities of AI to data collected by IoT, companies can identify and understand patterns and make more informed decisions. This leads to a variety of benefits for both consumers and companies such as proactive intervention, intelligent automation, and highly personalized experiences. It also enables us to find ways for connected devices to work better together and make these systems easier to use.

This, in turn, leads to even higher adoption rates. That is exactly why we need to improve the speed and accuracy of data analysis with AI in order to see IoT live up to its promise. Collecting data is one thing, but sorting, analyzing, and making sense of that data is a completely different thing. That is why it is essential to develop faster and more accurate AIs in order to keep up with the sheer volume of data being collected as IoT starts to penetrate almost all aspects of our lives.

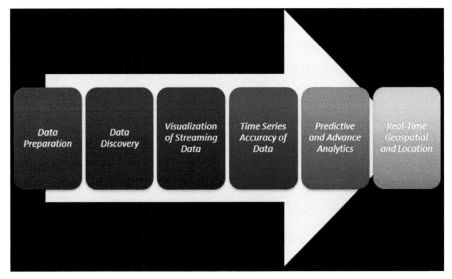

Figure 18.1   AI and IoT data analytics.

Examples of IoT Data [28]

- Data that helps cities predict accidents and crimes
- Data that gives doctors real-time insight into information from pacemakers or biochips
- Data that optimize productivity across industries through predictive maintenance on equipment and machinery
- Data that creates truly smart homes with connected appliances
- Data that provides critical communication between self-driving cars

It is simply impossible for humans to review and understand all of this data with traditional methods; even if they cut down the sample size, it simply takes too much time. The big problem will be finding ways to analyze the deluge of performance data and information that all these devices create. Finding insights in terabytes of machine data is a real challenge – just ask a data scientist.

But in order for us to harvest the full benefits of IoT data, we need to improve:

- *Speed* of big data analysis
- *Accuracy* of big data analysis

AI and IoT Data Analytics (Figure 18.1)

There are six types of IoT data analysis where AI can help [29]:

1. **Data preparation:** Defining pools of data and cleaning them, which will take us to concepts like dark data, data lakes, etc.
2. **Data discovery:** Finding useful data in the defined pools of data.
3. **Visualization of streaming data:** On-the-fly dealing with streaming data by defining, discovering, and visualizing data in smart ways to make it easy for the decision-making process to take place without delay.
4. **Time-series accuracy of data:** Keeping the level of confidence in data collected high with high accuracy and integrity of data.
5. **Predictive and advance analytics:** A very important step where decisions can be made based on data collected, discovered, and analyzed.
6. **Real-time geospatial and location (logistical data):** Maintaining the flow of data smooth and under control.

AI in IoT Applications [25]

- Visual big data, for example, will allow computers to gain a deeper understanding of images on the screen, with new AI applications that understand the context of images.
- Cognitive systems will create new recipes that appeal to the user's sense of taste, creating optimized menus for each individual, and automatically adapting to local ingredients.
- Newer sensors will allow computers to "hear" gathering sonic information about the user's environment.
- **Connected and remote operations:** With smart and connected warehouse operations, workers no longer have to roam the warehouse picking goods off the shelves to fulfill an order. Instead, shelves whisk down the aisles, guided by small robotic platforms that deliver the right inventory to the right place, avoiding collisions along the way. Order fulfillment is faster, safer, and more efficient.
- **Prevented/predictive maintenance:** Saving companies millions before any breakdown or leaks by predicting and preventing locations and time of such events.

These are just a few promising applications of artificial intelligence in IoT. The potential for highly individualized services are endless and will dramatically change the way people live.

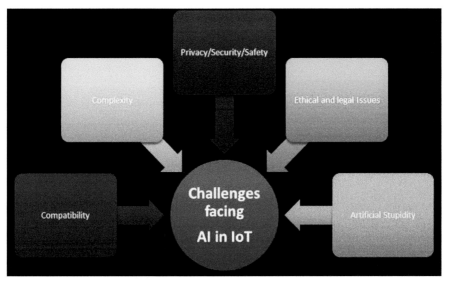

Figure 18.2   Challenges facing AI in IoT.

Challenges Facing AI in IoT (Figure 18.2)

1. **Compatibility:** IoT is a collection of many parts and systems that are fundamentally different in time and space.
2. **Complexity:** IoT is a complicated system with many moving parts and nonstop stream of data, making it a very complicated ecosystem.
3. **Privacy/security/safety (PSS):** Concerns are inherent with any emerging technology or concept. How can artificial intelligence (IA) be employed effectively to address these concerns without compromising PSS? One potential solution to this challenge is the adoption of blockchain technology.
4. **Ethical and legal issues:** It is a new world for many companies with no precedents, untested territory with new laws, and cases emerging rapidly.
5. **Artificial stupidity:** Back to the very simple concept of GIGO (garbage in garbage out), AI still needs "training" to understand human reactions/emotions so that the decisions will make sense.

Conclusion

While IoT is quite impressive, it really does not amount to much without a good AI system. Both technologies need to reach the same level of development in order to function as perfectly as we believe they should and would. Scientists are trying to find ways to make more intelligent data analysis software and devices in order to make safe and effective IoT a reality. It may take some time before this happens because AI development is lagging behind IoT, but the possibility is, nevertheless, there.

Integrating AI into IoT networks is becoming a prerequisite for success in today's IoT-based digital ecosystems. So businesses must move rapidly to identify how they will drive value from combining AI and IoT – or face playing catch-up in years to come.

The only way to keep up with this IoT-generated data and gain the hidden insights it holds is using AI as the *catalyst* of IoT.

19

Artificial Intelligence (AI) and Cloud Computing

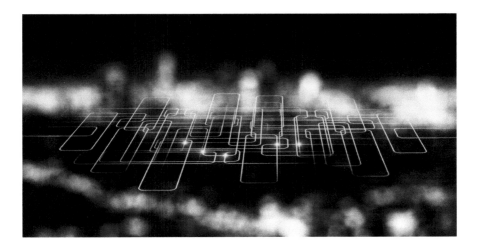

Artificial intelligence (AI) and cloud computing are two of the most rapidly evolving technologies in the world today. When combined, they offer endless possibilities for businesses and organizations to optimize their operations, gain insights from vast amounts of data, and transform the way they work. In this chapter, we will explore the challenges and opportunities of AI and cloud computing, as well as their future potential.

Advantages of using AI in Cloud Computing (Figure 19.1)

- **Scalability:** AI-powered cloud computing solutions allow businesses to easily scale up or down their computing resources based on their needs.

Advantages of using AI in Cloud Computing

•Scalability
•Automation
•Cost savings
•Enhanced security
•Improved customer experience

Disadvantages of using AI in Cloud Computing

• Dependency
• Lack of transparency
• Data privacy concerns
• Cost
• Skilled labor

Figure 19.1   Advantages and disadvantages of using AI in cloud computing.

- **Automation:** AI can automate many cloud computing tasks such as resource allocation, security monitoring, and performance optimization, freeing up IT staff to focus on more strategic initiatives.
- **Cost savings:** AI can help optimize resource allocation, reduce downtime, and increase efficiency, resulting in cost savings for businesses.
- **Enhanced security:** AI can be used to monitor and detect security threats in real time, reducing the risk of data breaches and cyberattacks.
- **Improved customer experience:** AI-powered cloud computing solutions can help businesses deliver personalized experiences to customers, such as chatbots and virtual assistants.

Disadvantages of using AI in Cloud Computing (Figure 19.1)

- **Dependency:** Businesses become more dependent on AI systems, and any errors or disruptions in the system can have significant consequences.

- **Lack of transparency:** AI algorithms can be complex and difficult to interpret, making it challenging for businesses to understand how decisions are made.
- **Data privacy concerns:** AI systems rely on vast amounts of data, which can raise privacy concerns if the data is not properly managed or secured.
- **Cost:** Implementing and maintaining AI systems can be expensive, particularly for small- and medium-sized businesses.
- **Skilled labor:** There is a shortage of skilled professionals who can develop, implement, and maintain AI systems.

Opportunities of using AI in Cloud Computing (Figure 19.2)

- **Improved decision-making:** AI can provide insights and recommendations to help businesses make better decisions.
- **Predictive analytics:** AI can analyze vast amounts of data to identify patterns and predict future trends, helping businesses stay ahead of the competition.
- **Personalization:** AI-powered solutions can help businesses deliver personalized experiences to customers, improving customer loyalty and satisfaction.

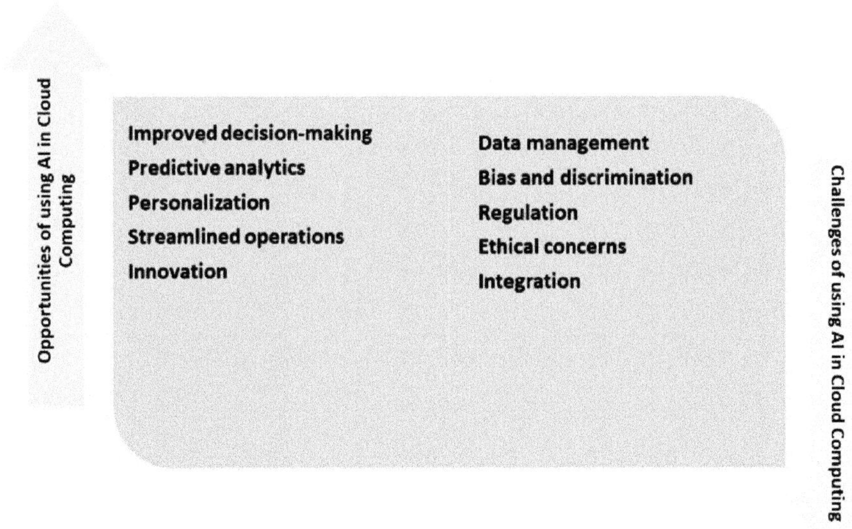

Figure 19.2   Opportunities and challenges of using AI in cloud computing.

- **Streamlined operations:** AI can automate many routine tasks, freeing up employees to focus on higher-value tasks.
- **Innovation:** AI can enable businesses to develop innovative new products and services, opening up new revenue streams.

Challenges of using AI in Cloud Computing (Figure 19.2)

- **Data management:** AI systems rely on vast amounts of data, which must be properly managed and secured.
- **Bias and discrimination:** AI algorithms can perpetuate bias and discrimination if the data used to train them is biased.
- **Regulation:** AI is a rapidly evolving field, and regulations may not keep pace with technological advancements, leading to uncertainty for businesses.
- **Ethical concerns:** AI raises a range of ethical concerns, such as the impact on jobs and privacy.
- **Integration:** Integrating AI with existing systems can be challenging, requiring significant time and resources.

Future of AI and Cloud Computing

The future of AI and cloud computing looks promising, with continued growth and innovation expected in both fields. As more businesses adopt cloud computing and AI, we can expect to see greater integration between these technologies, with AI services becoming an integral part of cloud computing platforms.

In addition, advancements in AI and cloud computing are likely to lead to new use cases and applications. For example, we may see the emergence of more sophisticated AI models that can analyze data in real time, predict outcomes, and automate decision-making processes. We may also see the development of more specialized cloud services, designed specifically for AI applications.

AI and cloud computing offer significant opportunities for businesses and organizations to drive innovation, improve efficiency, and enhance customer experiences. However, these technologies also pose significant challenges, particularly around data privacy and security, as well as the complexity of integrating new tools and systems into existing workflows.

Despite these challenges, the future of AI and cloud computing looks promising, with continued growth and innovation expected in both fields. As more businesses adopt these technologies, we can expect to see greater integration and specialization, as well as new use cases and applications that transform the way we work and live.

20

Your Computer Will Feel Your Pain

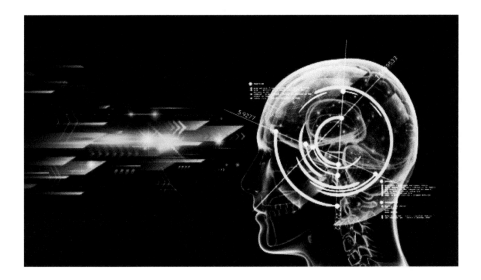

What if your smart device could empathize with you? The evolving field known as *affective computing* is likely to make it happen soon. Scientists and engineers are developing systems and devices that can recognize, interpret, process, and simulate human affects or emotions. It is an interdisciplinary field spanning computer science, psychology, and cognitive science. While its origins can be traced to longstanding philosophical inquiries into emotion, a 1995 paper on affective computing by Rosalind Picard catalyzed modern progress.

The more smart devices we have in our lives, the more we are going to want them to behave politely and be socially smart. We do not want them to bother us with unimportant information or overload us with too much information. That kind of common-sense reasoning requires an understanding of our emotional state. We are starting to see such systems perform specific, predefined functions, like changing in real time how you are presented with the questions in a quiz, or recommending a set of videos in an educational program to fit the changing mood of students.

How can we make a device that responds appropriately to your emotional state? Researchers are using sensors, microphones, and cameras combined with software logic. A device with the ability to detect and appropriately respond to a user's emotions and other stimuli could gather cues from a variety of sources. Facial expressions, posture, gestures, speech, the force or rhythm of key strokes, and the temperature changes of a hand on a mouse can all potentially signify emotional changes that can be detected and interpreted by a computer. A built-in camera, for example, may capture images of a user. Speech, gesture, and facial recognition technologies are being explored for affective computing applications.

Just looking at speech alone, a computer can observe innumerable variables that may indicate emotional reaction and variation. Among these are a person's rate of speaking, accent, pitch, pitch range, final lowering, stress frequency, breathlessness, brilliance, loudness, and discontinuities in the pattern of pauses or pitch.

Gestures can also be used to detect emotional states, especially when used in conjunction with speech and face recognition. Such gestures might include simple reflexive responses, like lifting your shoulders when you do not know the answer to a question. Or they could be complex and meaningful, as when communicating with sign language.

A third approach is the monitoring of physiological signs. These might include pulse and heart rate or minute contractions of facial muscles. Pulses in blood volume can be monitored, as can what is known as galvanic skin response. This area of research is still relatively new, but it is gaining momentum and we are starting to see real products that implement the techniques.

Recognizing emotional information requires the extraction of meaningful patterns from the gathered data. Some researchers are using machine learning techniques to detect such patterns.

Detecting emotion in people is one thing. But work is also going into computers that themselves show what appear to be emotions. Already in use are systems that simulate emotions in automated telephone and online conversation agents to facilitate interactivity between humans and machines.

There are many applications for affective computing. One is in education. Such systems can help address one of the major drawbacks of online learning versus in-classroom learning: the difficulty faced by teachers in adapting pedagogical situations to the emotional state of students in the classroom. In e-learning applications, affective computing can adjust the presentation style of a computerized tutor when a learner is bored, interested, frustrated, or pleased. Psychological health services also benefit from affective computing applications that can determine a client's emotional state.

Robotic systems capable of processing affective information can offer more functionality alongside human workers in uncertain or complex environments. Companion devices, such as digital pets, can use affective computing abilities to enhance realism and display a higher degree of autonomy.

Other potential applications can be found in social monitoring. For example, a car might monitor the emotion of all occupants and invoke additional safety measures, potentially alerting other vehicles if it detects the driver to be angry. Affective computing has potential applications in human–computer interaction, such as affective "mirrors" that allow the user to see how he or she performs. One example might be warning signals that tell a driver if they are sleepy or going too fast or too slow. A system might even call relatives if the driver is sick or drunk (though one can imagine mixed reactions on the part of the driver to such developments). Emotion-monitoring agents might issue a warning before one sends an angry email, or a music player could select tracks based on your mood. Companies may even be able to use affective computing to infer whether their products will be well-received by the market by detecting facial or speech changes in potential customers when they read an ad or first use the product. Affective computing is also starting to be applied to the development of communicative technologies for use by people with autism.

Many universities have done extensive work on affective computing resulting projects that include something called the galvactivator, which was a good starting point. It is a glove-like wearable device that senses a wearer's skin conductivity and maps values to a bright LED display. Increases in skin conductivity across the palm tend to indicate physiological arousal; so the display glows brightly. This may have many potentially useful purposes, including self-feedback for stress management, facilitation of conversation between two people, or visualizing aspects of attention while learning. Along with the revolution in wearable computing technology, affective computing is poised to become more widely accepted, and there will be endless applications for affective computing in many aspects of life.

One of the future applications will be the use of affective computing in Metaverse applications, which will humanize the avatar and add emotions as the fifth dimension opens limitless possibilities, but all these advancements in applications of affective computing racing to make the machines more human will come with challenges namely SSP (security, safety, privacy) – the three pillars of online users. We need to make sure that all the three pillars of online users are protected and well defined. It is easier said than done, but clear guidelines of what, where, and who will use the data will make acceptance of the hardware and software of affective computing faster without replacing the physical pain with the mental pain of fear of privacy, security, and safety of our data.

21

What is Autonomic Computing?

Autonomic computing is a computer's ability to manage itself automatically through adaptive technologies that further compute capabilities and cut down on the time required by computer professionals to resolve system difficulties and other maintenance such as software updates.

The move toward autonomic computing is driven by a desire for cost reduction and the need to lift the obstacles presented by computer system complexities to allow for more advanced computing technology.

The autonomic computing initiative (ACI), which was developed by IBM, demonstrates and advocates networking computer systems that do not involve a lot of human intervention other than defining input rules. The ACI is derived from the autonomic nervous system of the human body.

IBM has defined the four areas of automatic computing:

- Self-configuration
- Self-healing (error correction)
- Self-optimization (automatic resource control for optimal functioning)
- Self-protection (identification and protection from attacks in a proactive manner)

Characteristics that every autonomic computing system should have include *automation*, *adaptivity*, and *awareness*.

Autonomic computing (AC) was conceived to emulate the operations of the human body's nervous system. Much like the autonomic nervous system, which reacts to stimuli independently of conscious control, an autonomic computing environment operates with a significant degree of artificial intelligence while remaining imperceptible to users. Similar to how the human body autonomously regulates various functions, such as adjusting internal temperature, fluctuating breathing rates, and secreting hormones in response to stimuli, the autonomic computing environment seamlessly and organically responds to the data and inputs it collects.

IBM has set forth eight conditions that define an autonomic system:

1. The system must know itself in terms of what resources it has access to, what its capabilities and limitations are, and how and why it is connected to other systems.
2. The system must be able to automatically configure and reconfigure itself depending on the changing computing environment.
3. The system must be able to optimize its performance to ensure the most efficient computing process.
4. The system must be able to work around encountered problems by either repairing itself or routing functions away from the trouble.
5. The system must detect, identify, and protect itself against various types of attacks to maintain the overall system security and integrity.
6. The system must be able to adapt to its environment as it changes, interacting with neighboring systems and establishing communication protocols.
7. The system must rely on open standards and cannot exist in a proprietary environment.
8. The system must anticipate the demand on its resources while keeping transparent to users.

Autonomic computing is one of the building blocks of *pervasive computing*, an anticipated future computing model in which tiny — even invisible — computers will be all around us, communicating through increasingly interconnected networks leading to the concept of the Internet of Everything (IoE). Many industry leaders are researching various components of autonomic computing.

Benefits

The main benefit of autonomic computing is reduced TCO (total cost of ownership). Breakdowns will be less frequent, thereby drastically reducing maintenance costs. Fewer personnel will be required to manage the systems. "The most immediate benefit of autonomic computing will be reduced deployment and maintenance cost, time, and increased stability of IT systems through automation," says Dr. Kumar of IBM. "Higher order benefits will include allowing companies to better manage their business through IT systems that are able to adopt and implement directives based on business policy, and are able to make modifications based on changing environments."

Another benefit of this technology is that it provides server consolidation to maximize system availability and minimizes cost and human effort to manage large server farms.

Future of Autonomic Computing

Autonomic computing promises to simplify the management of computing systems. But that capability will provide the basis for much more effective cloud computing. Other applications include server load balancing, process allocation, monitoring power supply, automatic updating of software and drivers, pre-failure warning, memory error-correction, automated system backup and recovery, etc.

22

What is Affective Computing?

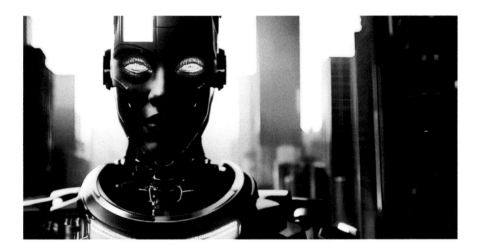

Affective computing is the study and development of systems and devices that can recognize, interpret, process, and simulate human affects. It is an interdisciplinary field spanning computer science, psychology, and cognitive science. While the origins of the field may be traced as far back as to early philosophical inquiries into emotion ("affect" is, basically, a synonym for "emotion"), the more modern branch of computer science originated with *Rosalind Picard's* 1995 paper on affective computing. A motivation for the research is the ability to simulate *empathy*. The machine should interpret the emotional state of humans and adapt its behavior to them, giving an appropriate response for those emotions.

Affective computing technologies sense the emotional state of a user (via sensors, microphone, cameras, and/or software logic) and respond by performing specific, predefined product/service features, such as changing a quiz or recommending a set of videos to fit the mood of the learner.

The more computers we have in our lives, the more we are going to want them to behave politely and be socially smart. We do not want it to bother us with unimportant information. That kind of common-sense reasoning requires an understanding of the person's *emotional state.*

One way to look at affective computing is *human−computer interaction* in which a device has the ability to detect and appropriately respond to its user's emotions and other stimuli. A computing device with this capacity could gather cues to user emotion from a variety of sources. Facial expressions, posture, gestures, speech, the force or rhythm of key strokes, and the temperature changes of the hand on a mouse can all signify changes in the user's emotional state, and these can all be detected and interpreted by a computer. A built-in camera captures images of the user, and algorithms are used to process the data to yield meaningful information. Speech recognition and gesture recognition are among the other technologies being explored for affective computing applications.

Recognizing emotional information requires the extraction of meaningful patterns from the gathered data. This is done using machine learning techniques that process different modalities, such as speech recognition, natural language processing, or facial expression detection.

Emotion in Machines

A major area in affective computing is the design of computational devices proposed to exhibit either innate emotional capabilities or that are capable of convincingly simulating emotions. A more practical approach, based on current technological capabilities, is the simulation of emotions in conversational agents in order to enrich and facilitate interactivity between humans and machines. While human emotions are often associated with surges in hormones and other neuropeptides, emotions in machines might be associated with abstract states associated with progress (or lack of progress) in autonomous learning systems. In this view, affective emotional states correspond to time derivatives in the learning curve of an arbitrary learning system.

Two major categories describing emotions in machines are: *emotional speech* and *facial affect detection.*

Emotional speech includes:

• Algorithms
• Databases

- Speech descriptors

Facial affect detection includes:

- Body gesture
- Physiological monitoring

The Future

Affective computing tries to address one of the major drawbacks of *online learning* versus in-classroom learning — the teacher's capability to immediately adapt the pedagogical situation to the emotional state of the student in the classroom. In e-learning applications, affective computing can be used to adjust the presentation style of a computerized tutor when a learner is bored, interested, frustrated, or pleased. *Psychological health services*, i.e., counseling, benefit from affective computing applications when determining a client's emotional state.

Robotic systems capable of processing affective information exhibit higher flexibility while one works in uncertain or complex environments. Companion devices, such as digital pets, use affective computing abilities to enhance realism and provide a higher degree of autonomy.

Other potential applications are centered around *social monitoring*. For example, a car can monitor the emotion of all occupants and engage in additional safety measures, such as alerting other vehicles if it detects the driver to be angry. Affective computing has potential applications in human—computer interactions, such as affective mirrors allowing the user to see how he or she performs; emotion monitoring agents sending a warning before one sends an angry email; or even music players selecting tracks based on mood. Companies would then be able to use affective computing to infer whether their products will or will not be well received by the respective market. There are endless applications for affective computing in all aspects of life.

23

Quantum Machine Learning (QML)

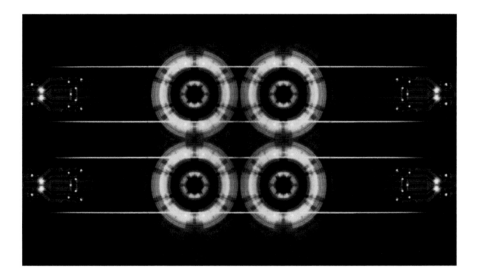

Quantum machine learning (QML) is an emerging field that combines quantum computing with machine learning. *Machine learning* is a subfield of artificial intelligence that focuses on developing algorithms that can learn from data and make predictions or decisions without being explicitly programmed. *Quantum computing* is the area of study focused on developing computer technology based on the principles of quantum theory.

The quantum computer, following the laws of quantum physics, would gain enormous processing power through the ability to be in multiple states, and to perform tasks using all possible permutations simultaneously.

A Comparison of Classical and Quantum Computing

Classical computing relies, at its ultimate level, on principles expressed by Boolean algebra. Data must be processed in an exclusive binary state at any point in time or bits. While the time that each transistor or capacitor need to be either in 0 or 1 before switching states is now measurable in billionths of a second, there is still a limit as to how quickly these devices can be made to switch states. As we progress to smaller and faster circuits, we begin to reach the physical limits of materials and the threshold for classical laws of physics to apply. Beyond this, the quantum world takes over. In a quantum computer, a number of elemental particles such as electrons or photons can be used with either their *charge* or *polarization* acting as a representation of 0 and/or 1. Each of these particles is known as a quantum bit, or *qubit*; the nature and behavior of these particles form the basis of quantum computing.

QML seeks to harness the power of quantum computing to improve machine learning algorithms and solve complex problems that classical computers cannot. In QML, quantum computing is used to perform operations on quantum data, which are represented by *quantum states*. These quantum states can encode information in a way that allows for more efficient processing and storage of data.

Components of QML (Figure 23.1)

The components of quantum machine learning include:

- **Quantum circuits:** Quantum circuits are the building blocks of quantum algorithms. They are a series of *quantum gates* that operate on *qubits* to perform calculations.
- **Quantum data:** Quantum data refers to data encoded in quantum states that can be manipulated by quantum algorithms. This data is typically represented as a collection of qubits.
- **Quantum algorithms:** Quantum algorithms are the algorithms that operate on quantum circuits to solve machine learning problems. These algorithms leverage the power of quantum computing to solve problems that are intractable for classical computers.
- **Quantum variational circuits:** Quantum variational circuits are a type of quantum circuit that can be *trained* to solve optimization problems using classical optimization techniques.

- **Quantum neural networks:** Quantum neural networks are a type of quantum circuit that can be trained to solve machine learning problems using *backpropagation.*
- **Quantum support vector machines:** Quantum support vector machines are a type of quantum algorithm that can be used to *classify* data into different categories.
- **Quantum principal component analysis:** Quantum principal component analysis is a quantum algorithm that can be used to reduce the *dimensionality* of large datasets.

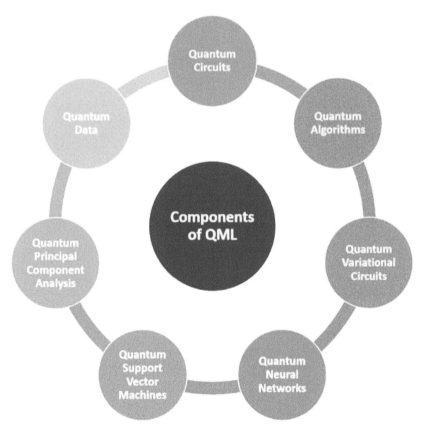

Figure 23.1   Components of QML.

Advantages of QML

One of the main advantages of QML is its ability to perform calculations on a large number of possible inputs simultaneously, a process known as *quantum parallelism*: Its ability to perform certain types of computations *exponentially* faster than classical computers. This is due to the fact that quantum computers can simultaneously compute many different outcomes at the same time. This can make it possible to solve problems that are currently intractable for classical computers.

Another advantage is the ability of quantum computing to perform *optimization* problems more efficiently, which is useful in fields such as logistics and finance. One more advantage of QML is that it can enable the development of new types of algorithms that are not possible with classical computers. For example, quantum machine learning algorithms could be used to perform computations that are not based on *classical probability distributions* or to create new models that can represent complex quantum states. QML has the potential to improve a wide range of machine learning tasks, such as *data clustering, classification,* and *regression analysis.* It can also be used for tasks such *as image and speech recognition, natural language processing,* and *recommendation systems.*

Another key advantage of QML is its ability to perform *unsupervised learning tasks* more efficiently. Unsupervised learning refers to the process of finding patterns in data without the use of labeled examples. This is an important area of machine learning, as it can be difficult and time-consuming to manually label large datasets.

QML could also have significant implications for fields such as finance, where it could be used to create more accurate models for predicting market trends and optimizing investment strategies.

Challenges Facing QML

QML is still in its early stages of development, and there are many challenges that need to be overcome before it can be widely adopted. These challenges include the need for better quantum hardware, improved algorithms, and better understanding of the relationship between quantum computing and machine learning. In addition to that, one of the main challenges is the difficulty of building stable and scalable quantum computers. At present, quantum computers are still in the early stages of development, and there are significant technical hurdles that need to be overcome before they can be widely adopted.

One of the main challenges in developing QML is the "*quantum-classical gap.*" This refers to the difficulty of translating classical machine learning algorithms into quantum algorithms, and vice versa. Researchers are working on developing new techniques to bridge this gap and create *hybrid algorithms* that take advantage of both classical and quantum computing.

There is currently a lot of interest in developing quantum machine learning algorithms that are "*quantum-inspired.*" These algorithms do not actually run on a quantum computer, but they are designed to take advantage of certain quantum properties to create more efficient machine learning models.

Despite these challenges, there is a great deal of interest in QML among researchers and practitioners in the field of machine learning. As more progress is made in developing stable and scalable quantum computers, and as new algorithms and techniques are developed, we can expect to see significant advances in QML over the coming years.

24

Applications of ChatGPT

ChatGPT, a large language model developed by OpenAI, has been making waves in the field of natural language processing (NLP) since its inception. The model has been trained on an enormous corpus of text and is capable of generating human-like responses to various prompts. ChatGPT has numerous applications in different fields, and in this essay, we will discuss some of the most promising ones. ChatGPT is a state-of-the-art language model that belongs to the category of deep learning models called *transformers*. It was trained using a variant of the transformer architecture known as the *generative pre-trained transformer* (GPT), which is a type of unsupervised learning method that uses large amounts of text data to train the model to generate coherent and relevant text.

In terms of AI classification, ChatGPT would be classified as an example of *artificial general intelligence (AGI)*, which is a type of AI that can perform intellectual tasks that are typically associated with humans, such as natural language processing and understanding, reasoning, decision-making, and problem-solving. However, it is important to note that ChatGPT is not a fully self-aware or conscious AI, but rather a sophisticated language processing tool.

One of the most exciting applications of ChatGPT is in the *customer service industry*. Many companies have already implemented chatbots on their websites to provide customer support, but these chatbots often lack the ability to understand and respond to complex queries. ChatGPT, on the other hand, can understand natural language and generate relevant responses based on the context of the conversation. This makes ChatGPT an ideal solution for customer support queries that require a human touch. For instance, if a customer has a question about a product's features or wants to report an issue, ChatGPT can respond with appropriate answers or direct the customer to the appropriate department.

Another promising application of ChatGPT is in the field of mental health. Mental health professionals are in high demand, but there is often a shortage of qualified practitioners. ChatGPT can provide a solution by acting as a virtual therapist. By asking relevant questions and generating responses based on the user's answers, ChatGPT can help users identify the underlying causes of their mental health issues and provide guidance on how to address them. This can be particularly beneficial for individuals who do not have access to traditional therapy services or are reluctant to seek them out.

ChatGPT can also be used to improve educational outcomes. By generating personalized study plans based on a student's learning style and strengths and weaknesses, ChatGPT can help students optimize their learning process. Additionally, ChatGPT can generate interactive exercises and quizzes that can help students reinforce their understanding of a particular subject. This can be particularly beneficial for students who struggle with traditional classroom instruction or who are unable to attend school due to geographic or financial constraints.

ChatGPT can also be used to improve online marketing efforts. By analyzing customer data and generating personalized recommendations based on their interests and preferences, ChatGPT can help businesses increase engagement and sales. Additionally, ChatGPT can generate engaging and personalized content that can help businesses build brand loyalty and improve their online presence.

In conclusion, ChatGPT has numerous applications in various fields, including customer service, mental health, education, and online marketing. As the technology continues to improve, we can expect to see more innovative applications of ChatGPT in the future. While there are some concerns about the ethical implications of using ChatGPT, such as privacy concerns and the potential for bias in the model, these issues can be addressed through proper implementation and regulation. Overall, ChatGPT is a powerful tool that has the potential to revolutionize the way we interact with technology and each other.

ChatGPT has been trained on an enormous corpus of text, including books, websites, and other written materials. The model uses a technique called transformer architecture, which allows it to understand and generate responses based on the context of the conversation. This means that ChatGPT can not only understand what the user is saying but also provide relevant responses based on the conversation's history and context.

One of the most exciting things about ChatGPT is its ability to generate natural-sounding responses. Unlike traditional chatbots, which often rely on pre-written scripts, ChatGPT can generate responses that sound like they were written by a human. This is because ChatGPT has been trained on a vast amount of text and can recognize and replicate the patterns and nuances of human language.

In addition to the applications mentioned in the previous section, Chat-GPT has many other potential applications. For example, ChatGPT can be used to assist with language translation, allowing users to communicate with people who speak different languages without the need for a human translator. Additionally, ChatGPT can be used to assist with research, helping researchers to identify relevant literature and generate insights based on large amounts of data.

There are, however, some concerns about the ethical implications of using ChatGPT. One of the most significant concerns is the potential for bias in the model. Because ChatGPT has been trained on a large corpus of text, it may inadvertently replicate biases and stereotypes present in the training data. This means that the model may be more likely to generate responses that are discriminatory or offensive.

Another concern is privacy. ChatGPT requires access to large amounts of data to function, which means that users' conversations may be recorded and analyzed. This raises questions about who has access to this data and how it is being used.

Despite these concerns, there is no denying that ChatGPT has enormous potential to improve our lives in many ways. As the technology continues to develop, we can expect to see more innovative applications of ChatGPT in fields such as healthcare, education, and entertainment. With proper regulation and oversight, ChatGPT could be a powerful tool for improving human communication and understanding.

Another exciting application of ChatGPT is in the field of journalism. ChatGPT can generate news articles based on a given topic, allowing news organizations to quickly produce content and stay up-to-date with breaking news. This can be particularly useful for smaller news organizations or those with limited resources, as ChatGPT can generate articles quickly and efficiently.

ChatGPT can also be used to assist with legal research. The model can analyze large amounts of legal text, including case law and statutes, and generate summaries or answers to specific legal questions. This can be a valuable tool for lawyers, judges, and other legal professionals, as it can save time and provide quick access to relevant information.

Another potential application of ChatGPT is in the field of creative writing. The model can generate prompts or story ideas, provide feedback on writing samples, and even generate entire stories or novels. This can be particularly useful for aspiring writers who are looking for inspiration or feedback on their work.

ChatGPT can also be used to improve accessibility for people with disabilities. For example, the model can generate descriptions of images or videos for people with visual impairments or provide real-time captioning for people with hearing impairments. This can help to make the internet and other digital media more accessible and inclusive for people with disabilities.

Finally, ChatGPT can be used to improve scientific research. The model can analyze scientific texts, generate summaries or abstracts, and even assist with scientific writing. This can be particularly useful for researchers who are looking to stay up-to-date with the latest research in their field or need assistance with writing and publishing their work.

Overall, ChatGPT has numerous potential applications in a wide range of fields. While there are certainly concerns about the ethical implications of using the model, such as privacy and bias, these issues can be addressed through proper implementation and regulation. As the technology contin-ues to develop, we can expect to see even more innovative applications of ChatGPT in the future.

25

AI and the Future of Work

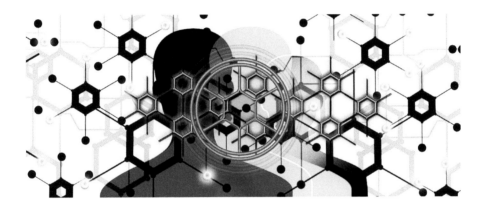

Artificial intelligence (AI) is a rapidly growing field that has the potential to revolutionize the way we work, learn, and interact with technology. The term AI refers to the ability of machines to perform tasks that would typically require human intelligence, such as decision-making, problem-solving, and natural language processing. As AI technology continues to advance, it is becoming increasingly integrated into various aspects of the workplace, from automating repetitive tasks to helping professionals make more informed decisions.

The impact of AI on the future of work is a topic of much discussion and debate. Some experts believe that AI will lead to the displacement of human workers, while others argue that it will create new opportunities and lead to increased productivity and economic growth. Regardless of the outcome, it is clear that AI will have a profound effect on the job market and the skills needed to succeed in the workforce.

In this context, it is crucial to understand the potential benefits and risks of AI in the workplace, as well as the ethical implications of using AI to make decisions that affect human lives. As AI continues to evolve, it is essential that both individuals and organizations stay informed and adapt to the changing landscape of work.

AI is set to transform the future of work in a number of ways. Following are some possible angles:

- **The impact of AI on jobs:** One of the biggest questions surrounding AI and the future of work is what impact it will have on employment. Will AI create new jobs or displace existing ones? What types of jobs are most likely to be affected?
- **The role of AI in workforce development:** As AI becomes more prevalent in the workplace, it is likely that workers will need to develop new skills in order to keep up. How can companies and organizations help workers develop these skills?
- **The future of collaboration between humans and AI:** Many experts believe that the future of work will involve collaboration between humans and AI. What might this collaboration look like? How can companies and organizations foster effective collaboration between humans and AI?
- **AI and workforce diversity:** AI has the potential to reduce bias and increase diversity in the workplace. How can organizations leverage AI to improve workforce diversity?
- **The ethical implications of AI in the workplace:** As AI becomes more prevalent in the workplace, there are a number of ethical considerations that need to be taken into account. How can companies and organizations ensure that their use of AI is ethical and responsible?
- **AI and the gig economy:** AI has the potential to transform the gig economy by making it easier for individuals to find work and for companies to find workers. How might AI impact the future of the gig economy?
- **AI and workplace automation:** AI is likely to automate many routine tasks in the workplace, freeing up workers to focus on higher-level tasks. What types of tasks are most likely to be automated, and how might this change the nature of work?

Advantages and disadvantages of AI in the context of the future of work (Figure 25.1):

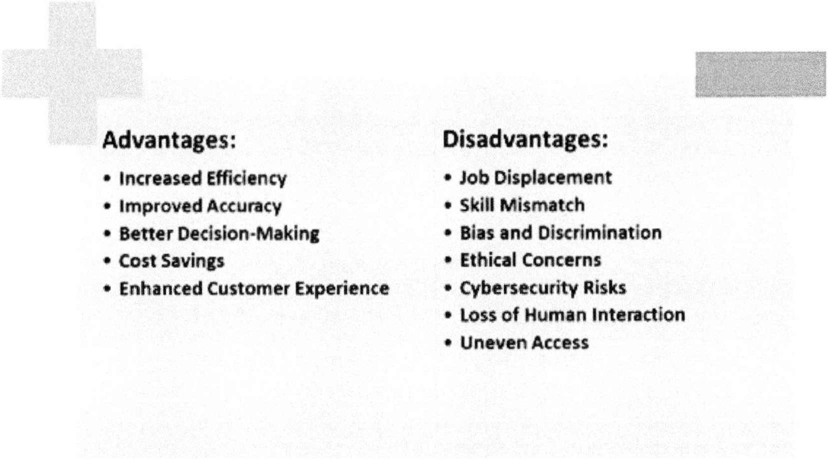

Advantages:	Disadvantages:
• Increased Efficiency	• Job Displacement
• Improved Accuracy	• Skill Mismatch
• Better Decision-Making	• Bias and Discrimination
• Cost Savings	• Ethical Concerns
• Enhanced Customer Experience	• Cybersecurity Risks
	• Loss of Human Interaction
	• Uneven Access

Figure 25.1   Advantages and disadvantages of AI in the context of the future of work.

Advantages:

- **Increased efficiency:** AI can automate many routine tasks and work-flows, freeing up workers to focus on higher-level tasks and increasing productivity.
- **Improved accuracy:** AI systems can process large amounts of data quickly and accurately, reducing the risk of errors.
- **Better decision-making:** AI can analyze data and provide insights that humans may not be able to identify, leading to better decision-making.
- **Cost savings:** By automating tasks and workflows, AI can reduce labor costs and improve the bottom line for businesses.
- **Enhanced customer experience:** AI-powered chatbots and other tools can provide fast, personalized service to customers, improving their overall experience with a company.

Disadvantages:

- **Job displacement:** As mentioned earlier, AI and automation could displace many workers, particularly those in low-skill jobs.
- **Skill mismatch:** As AI and automation become more prevalent, workers will need to develop new skills in order to remain competitive in the workforce.

- **Bias and discrimination:** AI systems are only as unbiased as the data they are trained on, which could lead to discrimination in hiring, promotion, and other workplace practices.
- **Ethical concerns:** As AI and automation become more prevalent, there are a number of ethical concerns that need to be addressed, including issues related to privacy, transparency, and accountability.
- **Cybersecurity risks:** As more and more data is collected and processed by AI systems, there is a risk that this data could be compromised by cybercriminals.
- **Loss of human interaction:** AI systems may replace some forms of human interaction in the workplace, potentially leading to a loss of social connections and collaboration between workers.
- **Uneven access:** As mentioned earlier, not all workers and organizations have equal access to AI and automation technology, which could widen the gap between those who have access to these tools and those who do not.

These are just a few of the advantages and disadvantages of AI and the future of work. As AI continues to evolve, it is likely that new advantages and disadvantages will emerge as well.

In conclusion, the impact of AI on the future of work is a complex and multifaceted issue that requires careful consideration and planning. While AI has the potential to revolutionize the way we work and improve productivity, it also poses significant challenges, including job displacement and ethical concerns.

To prepare for the future of work, individuals and organizations must *prioritize upskilling and reskilling* to ensure that they have the skills and knowledge necessary to thrive in an AI-driven world. Additionally, policymakers must address the potential impacts of AI on employment and work toward creating policies that ensure the benefits of AI are shared equitably.

Ultimately, the successful integration of AI into the workplace will require collaboration and dialogue between industry, academia, and government to ensure that AI is used in a way that benefits society as a whole. By staying informed and proactive, we can navigate the changes brought about by AI and create a future of work that is both efficient and equitable.

26

New AI, New Jobs!

Artificial intelligence (AI) is a rapidly evolving field that is changing the way we live, work, and interact with the world. One of the most significant impacts of AI has been on the *job market*. While there is much concern about the loss of jobs due to automation, there are also *many new jobs being created* in AI and related fields.

- **Prompt engineering:** It is a field of work that involves developing natural language processing (NLP) models and algorithms that can understand and respond to human language inputs in a way that is accurate, relevant, and coherent. This is the core technology behind many AI-powered systems that use natural language as an interface, such as chatbots, virtual assistants, and voice-controlled devices.
- **AI ethics and governance specialist:** As AI systems become more prevalent, there is a growing need for professionals who can ensure that

these systems are developed and deployed ethically and responsibly. AI ethics and governance specialists help organizations to navigate complex ethical and legal issues related to AI, including issues around privacy, bias, and transparency.

- **AI business strategist:** As more companies invest in AI, there is a growing need for professionals who can help them to develop and execute AI strategies that align with their business goals. AI business strategists work with executives and stakeholders to identify opportunities for AI, develop business cases for AI investments, and ensure that AI initiatives are aligned with broader business strategies.

- **AI product manager:** AI product managers are responsible for developing and managing AI-powered products and services. They work with cross-functional teams to define product requirements, develop roadmaps, and ensure that products are delivered on time and within budget. They also need to stay up-to-date with the latest developments in AI technology and market trends to ensure that their products remain competitive.

- **AI user experience designer:** As AI becomes more ubiquitous, there is a growing need for professionals who can design user experiences that are intuitive, engaging, and responsive. AI user experience designers work with product teams to create user interfaces that are optimized for AI-powered interactions, such as chatbots, virtual assistants, and recommendation engines.

- **AI cybersecurity analyst:** As AI systems become more complex, they also become more vulnerable to cyberattacks. AI cybersecurity analysts help organizations to identify and mitigate security risks related to AI, including issues related to data privacy, system vulnerabilities, and insider threats.

- **AI engineers**, for example, are responsible for creating and integrating AI algorithms into existing systems. They need to understand both the technical aspects of AI and the business needs of the organization they work for. They also need to stay up-to-date with the latest developments in AI technology to ensure that their systems remain competitive.

- **Data scientists and machine learning specialists**, on the other hand, work with large datasets to identify patterns and develop predictive models. They use algorithms to train AI systems to recognize and respond to specific inputs, such as images, speech, or text. This requires a deep understanding of statistics, mathematics, and computer science, as well as domain-specific knowledge in fields such as medicine, finance, or engineering.

- **Software developers** are responsible for creating the applications and platforms that run AI systems. They use programming languages such as Python, Java, or C++ to write the code that enables AI systems to communicate, learn, and make decisions. They also need to work closely with other professionals to ensure that the software is secure, scalable, and user-friendly.

These are just a few examples of the many new roles that are emerging in the AI job market. As AI continues to evolve, we can expect to see even more opportunities for workers with a range of skills and expertise. While these roles are critical to the development of AI systems, there are also many other jobs being created in related fields. AI is driving innovation in the *healthcare* industry, leading to the creation of new jobs in fields such as telemedicine, medical imaging, and personalized medicine. AI-powered tools are helping doctors and nurses diagnose and treat patients more accurately and efficiently and are also facilitating medical research by analyzing vast amounts of data.

Another area where AI is creating jobs is in *customer service* and sales. Chatbots, virtual assistants, and other AI-powered tools are increasingly being used to interact with customers, answer questions, and provide personalized recommendations. This requires a range of skills, including natural language processing, customer service, and sales.

Beyond these specific roles, AI is also driving the growth of new industries and business models. For example, AI is fueling the rise of the *gig economy*, as platforms such as Uber, Lyft, and Airbnb use AI to match workers with jobs and customers with services. Similarly, AI is driving the growth of *e-commerce*, as companies such as Amazon and Alibaba use AI to personalize recommendations, optimize logistics, and improve customer experiences.

Indubitably, the rise of AI also presents some challenges. One of the most significant concerns is the potential for *job displacement*, as AI systems automate tasks that were previously performed by humans. This could lead to widespread unemployment in some industries, particularly those that rely on manual labor or routine tasks.

To address this challenge, *policymakers and business leaders* will need to focus on retraining and reskilling programs to help workers transition into new roles. They will also need to promote the development of new industries and business models that can create new jobs and provide new opportunities for workers.

While AI has created new job opportunities, it has also brought about certain *challenges* that must be addressed. Some of the challenges facing the job market due to AI advancements are:

1. **Automation:** The increasing use of AI in the job market has led to concerns about job displacement, particularly in industries that rely heavily on manual labor or repetitive tasks. While AI has the potential to increase productivity and efficiency, it can also lead to *job losses* if workers are replaced by machines.

2. **Skill mismatch:** The use of AI in the job market has also led to a shift in the types of skills that are in demand. Technical skills such as coding, data analysis, and machine learning are increasingly in demand, while some traditional skills such as manual labor and clerical work may be less in demand. This has led to a skill mismatch, where workers may not have the necessary skills to succeed in the job market.

3. **Bias:** One of the challenges facing the job market due to AI is the potential for bias. AI systems are only as unbiased as the data they are trained on. If the data used to train an AI system is biased, the system will reflect that bias. This can lead to discrimination in the job market, particularly in the recruitment and hiring process. For example, if an AI system is trained on data that is biased against certain ethnic or gender groups, it may result in discrimination against those groups in the *hiring process*.

4. **Privacy concerns:** The use of AI in the job market has raised concerns about the collection and use of personal data. Employers may collect data on job applicants through online profiles, social media, and other sources. This data may be used to screen candidates and make hiring decisions. There are concerns about how this data is collected, stored, and used, and whether it is being used fairly.

5. **Unequal distribution of benefits:** While AI has the potential to create new job opportunities and improve productivity, the benefits of these advancements may not be distributed equally. Some workers may benefit more than others, depending on their skills and job requirements. For example, workers with technical skills may benefit from the increasing demand for these skills, while workers without these skills may be left behind. Similarly, workers in certain industries may benefit more from AI than workers in other industries.

Generally, the challenges facing the job market due to AI advancements require careful *consideration and planning* to ensure that the benefits of AI are shared fairly and that workers are not left behind. This may involve *investing in education and training programs* to help workers acquire the necessary skills to succeed in the job market, ensuring that AI systems are transparent and unbiased, and addressing concerns about the collection and use of personal data.

References

[1] https://www.mygreatlearning.com/blog/what-is-artificial-intelligence/
[2] http://www.technologyreview.com/news/524026/is-google-cornering-the-market-on-deep-learning/
[3] http://www.forbes.com/sites/netapp/2013/08/19/what-is-deep-learning/
[4] http://www.fastcolabs.com/3026423/why-google-is-investing-in-deep-learning
[5] http://www.npr.org/blogs/alltechconsidered/2014/02/20/280232074/deep-learning-teaching-computers-to-tell-things-apart
[6] http://www.technologyreview.com/news/519411/facebook-launches-advanced-ai-effort-to-find-meaning-in-your-posts/
[7] http://www.deeplearning.net/tutorial/
[8] http://searchnetworking.techtarget.com/definition/neural-network
[9] https://en.wikipedia.org/wiki/Affective_computing
[10] http://www.gartner.com/it-glossary/affective-computing
[11] http://whatis.techtarget.com/definition/affective-computing
[12] http://curiosity.discovery.com/question/what-is-affective-computing
[13] https://www.linkedin.com/pulse/quantum-computing-blockchain-facts-myths-ahmed-banafa/
[14] https://analyticsindiamag.com/will-quantum-computing-define-the-future-of-ai/
[15] https://www.analyticsinsight.net/ai-quantum-computing-can-enable-much-anticipated-advancements/
[16] https://research.aimultiple.com/quantum-ai/
[17] https://www.globenewswire.com/news-release/2020/11/17/2128495/0/en/Quantum-Computing-Market-is-Expected-to-Reach-2-2-Billion-by-2026.html
[18] https://ai.googleblog.com/2019/10/quantum-supremacy-using-programmable.html
[19] https://aibusiness.com/ai-brain-iot-body/

[20] https://thenextweb.com/hardfork/2019/02/05/blockchain-and-ai-could
-be-a-perfect-match-heres-why/

[21] https://www.forbes.com/sites/darrynpollock/2018/11/30/the-fourth-ind
ustrial-revolution-built-on-blockchain-and-advanced-with-ai/#4cb2e
5d24242

[22] https://www.forbes.com/sites/rachelwolfson/2018/11/20/diversifying-d
ata-with-artificial-intelligence-and-blockchain-technology/#1572eefd4
dad

[23] https://hackernoon.com/artificial-intelligence-blockchain-passive-inco
me-forever-edad8c27844e

[24] https://blog.goodaudience.com/blockchain-and-artificial-intelligence-t
he-benefits-of-the-decentralized-ai-60b91d75917b

[25] https://aibusiness.com/ai-brain-iot-body/

[26] http://www.creativevirtual.com/artificial-intelligence-the-internet-of-t
hings-and-business-disruption/

[27] https://www.computer.org/web/sensing-iot/contentg=53926943&type=
article&urlTitle=what-are-the-components-of-iot-

[28] https://www.bbvaopenmind.com/en/the-last-mile-of-iot-artificial-intell
igence-ai/

[29] http://www.datawatch.com/

[30] https://www.pwc.es/es/publicaciones/digital/pwc-ai-and-iot.pdf

[31] http://www.iamwire.com/2017/01/iot-ai/148265

Index

About the Author

Prof. Ahmed Banafa is a distinguished expert in IoT, Blockchain, Cyber-security, and AI with a strong background in research, operations, and management. He has been recognized for his outstanding contributions, receiving the Certificate of Honor from the City and County of San Francisco, the Haskell Award for Distinguished Teaching from the University of Massachusetts Lowell, and the Author & Artist Award from San Jose State University. LinkedIn named him the No.1 tech voice to follow in 2018, acknowledging his predictive insights and influence. His groundbreaking research has been featured in renowned publications like Forbes, IEEE, and the MIT Technology Review. He has been interviewed by major media outlets including ABC, CBS, NBC, CNN, BBC, NPR, NHK, and FOX. Being a member of the MIT Technology Review Global Panel further highlights his prominence in the tech community.

Prof. Banafa is an accomplished author known for impactful books. His work "Secure and Smart Internet of Things (IoT) using Blockchain and Artificial Intelligence (AI)" earned him the San Jose State University Author and Artist Award and recognition as one of the Best Technology Books of All Time and Best AI Models Books of All Time. His book on "Blockchain Technology and Applications" also received acclaim and is integrated into curricula at prestigious institutions like Stanford University. Additionally, he has contributed significantly to Quantum Computing through his third book, and he is preparing to release his fourth book on Artificial Intelligence in 2023. Prof. Banafa's educational journey includes Cybersecurity studies at Harvard University and Digital Transformation studies at the Massachusetts Institute of Technology (MIT), he is holding a Master's Degree in Electrical Engineering and a PhD in Artificial Intelligence.